Studies of Brain Function, Vol. 3

Coordinating Editor
V. Braitenberg, Tübingen

Editors
H. B. Barlow, Cambridge
E. Bizzi, Cambridge, USA
E. Florey, Konstanz
O.-J. Grüsser, Berlin-West
H. van der Loos, Lausanne

J. T. Enright

The Timing of
Sleep and Wakefulness

On the Substructure and Dynamics of the Circadian
Pacemakers Underlying the Wake-Sleep Cycle

With a Foreword by E. Florey
and an Appendix by J. Thorson

With 103 Figures

Springer-Verlag
Berlin Heidelberg New York 1980

Professor JAMES T. ENRIGHT
University of California
Scripps Institution of Oceanography
La Jolla. CA 92093, USA

546984

QP
425
.E57
1980

ISBN 3-540-09667-1 Springer-Verlag Berlin Heidelberg New York
ISBN 0-387-09667-1 Springer-Verlag New York Heidelberg Berlin

Library of Congress Cataloging in Publication Data. Enright, James Thomas,
1932- The timing of sleep and wakefulness. (Studies of brain function ;
Bibliography: p. Includes index. 1. Sleep-Physiological aspects. 2. Circadian
rhythms--Mathematical models. 3. Circadian rhythms--Data processing.
4. Action potentials (Electrophysiology) I. Title. II. Series. QP425.E57
612'.821'0184 79-22480

© by Springer-Verlag Berlin Heidelberg 1980.
Printed in Germany.

Offsetprinting and binding: Julius Beltz, Hemsbach/Bergstr.
2131/3130-543210

Dedicated to Professor Jürgen Aschoff
with warm affection and deep respect

Foreword

The brain functions like a computer composed of subsystems which interact in a hierarchical manner. But it is not a single hierarchy, but a complex system of hierarchies each of which has its very own and unique feature. One of these concerns the cyclic or rhythmic control of neuronal activities which, among others, give rise to alternating states of wakefulness and sleep.

The phenomenon of sleep still remains a mystery. The present monograph does not give us any new insights into its meaning and significance. Yet sleep research may not be the same after the appearance of this book because it gives us a comprehensive mathematical theory which opens our eyes to new insights into the mechanism of the rhythm generation that underlies the "wake-sleep" cycle.

No one who has worked his way through this book can again look at experimental data without recognizing features which the "models" developed in its various chapters so strikingly reveal.

The importance of this new vista lies not only in the possibilities it generates for computer simulation of the effects of timed stimuli on parameters of diurnal rhythms; the new ideas presented here also indicate novel experiments — and, more important even, they force the physiologist to have a new look at the nervous system. Physiologists and behaviorists too often have tacitly assumed that the rhythm generator, the "clock", is a singular locus, possibly a single cell — the mysterious "oscillator" — which keeps time and controls the timing of rhythmic behavior.

Enright's interactionist view which envisions the participation of hundreds, thousands, possibly millions of neurons in rhythm generation may not be fully applicable to invertebrates with small-number nervous systems like those of gastropod molluscs. His concept cannot be overlooked, however, and has important implications even for the interpretations of rhythmic behavior of invertebrate species.

And who knows: these numerous elements (Enright's "pacers") that interact to produce precise pacemaker activity might not always be whole neurons but elementary subsystems within neurons — particularly if these are large. And if this were so, Enright's "ensemble" might well turn out to be present within a single cell after all.

While up to now there developed a growing confidence, that a "biological clock" can actually be found, Enright now shows us how it might work, — in fact the success of his approach, so clearly made evident in this book, suggests that his theory demonstrates how this clock ought to work.

Konstanz, April 1979   Ernst Florey

Preface

Twenty years ago, the daily rhythmicity of plants and animals was regarded by most biologists as an abstruse and purely descriptive area of research, important perhaps to ecologists but with little relevance to the broader concerns of modern experimental science. Today, however, the situation is very different. The attempts to understand these rhythms in terms of their physiological origins and consequences constitute one of the most rapidly expanding branches of biology. An extensive literature, based on a huge body of data, has been published. The basic phenomena have been rechristened, and are now known as "circadian" (= about a day) rhythms. The experimental methods employed range from simple manipulation of external stimuli to microsurgical intervention, from the techniques of genetics and biochemistry to those of the neurophysiologist. Today even the educated layman can speak knowingly of some of the consequences of his "biological clock".

In humans, and in other higher vertebrates as well, a daily interval of sleep is the most conspicuous manifestation of this clock. It is now generally recognized that the regular alternation between the waking and sleeping states is an overt expression of an endogenous circadian rhythm, and the general objective of this book is to explore certain hypotheses about the physiological mechanisms which underlie the wake-sleep rhythm. The quantitative consequences of these hypotheses will be examined by means of formal models, in which postulated physiological variables will be represented by mathematical symbols, and computer simulation of the behavior of the models is a central component of the research method described here.

The treatment of the subject matter, however, is not aimed solely nor even primarily at specialists in circadian rhythms, at advanced researchers in the study of sleep, or at those well versed in the use of mathematical models. Instead, the presentation is intended for a broader audience, in the belief that this case study has heuristic value for beginning graduate students in the neurosciences, experimental psychology and related disciplines. No prior background in the literature of circadian rhythms is presumed, no extensive mathematical preparation, no familiarity with the strategy of modeling or computer simulation. The reader with such specialized background should be prepared, therefore, to skip through familiar materials, concepts and caveats, which are included here for the outsider. There is, I think, substance here for the specialist as well, but a variety of issues of less general interest have been relegated to appendixes which the general reader is, then, encouraged to bypass.

The interpretation proposed here for the wake-sleep cycle has as its foundation the assumption that single nerve cells, acting independently, can internally generate oscillations with more or less circadian frequency. This assumption is simply an extrapolation of the experimental observation that many kinds of unicellular organisms are able to show circadian variation in their properties; but in taking that experimental fact as a point of departure, I have sidestepped one of the primary questions which arises from circadian research: how is it possible that a single cell can generate such long-period oscillations? The biochemical and biophysical origins of circadian periodicity constitute the focus of research in many laboratories today, and the whole armamentarium of molecular biology has been unsheathed in that research.

Eventually, such studies will lead to an understanding of the mechanistic details of single-cell circadian rhythms; but there is a large conceptual gap between elucidation of rhythms at that level, and the full understanding of the circadian performance of multicellular animals. There are many complex phenomena which are characteristic of the rhythms of higher vertebrates, and which are unlikely to be explainable as self-evident consequences of single-cell rhythms. For example, the whole-animal rhythm is often extremely precise in its timing; the rhythm can be "molded" in frequency by past experience; it shows complex responses to synchronizing stimuli, to phasic as well as to tonic sensory input; and it controls response systems such as the sleep-wake cycle which have no counterpart in the single cell. In order to account fully for these whole-animal responses, an intervening level of explanation will be required, which deals with the principles by which single-cell oscillations and their output are organized into a unified whole: how the cellular oscillations are affected by sensory input, how they interact with each other, and how they determine the behavior of the whole animal. These issues represent the level of explanation with which the present treatise deals.

More explicitly, the question to be examined here is how nerve cells — which are presumed to have an intrinsic circadian capacity — might be interconnected and might interact to produce a unitary oscillatory output such as that seen in the wake-sleep cycle. The consideration of this issue leads to the formulation of an explicit model to describe the behavior of a particular sort of simple neuronal network, in which random (stochastic) events play a major role. While detailed information about the biochemical and ionic events which underlie the circadian rhythms of the single cells is not necessary for the models, a formal characterization of the processes *is* required. For that purpose, I invoke data from nerve cells which show "tonic activity" in the high-frequency domain: a more or less rhythmic, steady-state discharge pattern in the absence of phasic stimulation. From the study of these endogenous oscillations, which have periods of a few milliseconds to a few minutes, and which can be accelerated or slowed by the intensity of tonic stimuli, a number of researchers have all been led to the same general sort of formal model, which can satisfactorily account for the observed discharge patterns. I propose to extrapolate these models for high-frequency neuronal rhythms to the circadian time domain by using an equivalent formalism to describe the hypothesized circadian oscillations in single nerve cells.

The initial motivation for this study was to find a simple plausible explanation for the remarkable temporal precision of the circadian wake-sleep rhythms of higher vertebrates: how can biological clocks be so precise? As an emergent, and, in many ways, unanticipated consequence of the proposed mode of coupling within an ensemble of neuronal oscillators, models of this sort can also account naturally for a large body of experimental data on other properties of the circadian wake-sleep rhythm, and on the influences of light on the overt cycle. In its narrowest context, then, this book describes an attempt to bridge the gap which presently separates neurophysiology from the experimental study of the circadian rhythms of higher vertebrates. In a broader context, however, it represents an example of a general line of investigation which I think will be of considerable importance in the future of neurophysiology and the study of behavior: the computer-assisted study of significant behavioral phenomena by means of simulation models which are based upon known or reasonably inferred properties of the nervous system.

The need for approaches of this sort arises because of serious practical limitations on what a neurophysiologist can, at present, measure directly. A variety of data are now available about possible consequences of single neuronal connections, and these data can be used to synthesize plausible explanations for elementary kinds of behavior, in terms of a small circuit diagram, provided that the behavioral event of interest is not far removed in complexity from a simple reflex. In the vertebrate brain, however, there are huge arrays of cells which are similar to each other in both morphology and behavior, ensembles of units which are by no means redundant, but which instead constitute an interconnected and interacting whole. Such arrays of cells are probably responsible for many interesting kinds of complex behavior, ranging from finely controlled muscle movements to decision making, from sleeping to learning. What are the likely consequences of elementary kinds of interconnections and interactions among these arrays of neurons which constitute the central nervous system? To what extent can such interconnections, now understood at an elementary level, also account for details of the behavior of the whole organisms?

Direct simultaneous and independent measurements from large arrays of neurons are not yet available, and probably will not be feasible for some time to come. Intuition cannot be relied upon as a guide to answers to the questions of interest − as I have often found to my dismay in the studies described here. Computer simulation offers an alternative means − and perhaps the only one possible now − for exploring some of the consequences of interactions within such large networks, in the search for answers to questions which otherwise may seem completely intractable. The nonspecialist reader will, I hope, find this book a useful introduction to some of the methods, the limitations and the possible accomplishments of computer-simulation models as a means of research in the behavioral sciences.

The final chapter of this book represents a summary which is intended to be comprehensible without reliance on detailed concepts developed in the text. For the reader who seeks a brief account of the substance of the treatise, therefore, that chapter can be read first, or perhaps better, after completion of Chap. 1, which is intended to set the stage for the more complex treatment which follows.

Acknowledgements

The ideas underlying this treatise can be clearly traced back to my long, pleasant and fruitful associations in the laboratories of Professor Jürgen Aschoff, of the Max-Planck-Institut für Verhaltensphysiologie. My colleagues there, particularly Drs. Aschoff, Hoffmann and Wever, and also Drs. Daan, Gwinner, Pohl and others, have contributed importantly not only to my thoughts about biological clocks, but also to my entire view of the business of science. The development of the computer models, as well as the collection of some of the data presented here, have been supported by a series of grants from the National Science Foundation and the National Institutes of Health. Dr. T.H. Bullock has played a major role in this undertaking by his interest and encouragement. Many others of my colleagues and acquaintances have helped shape my ideas by the kinds of experimental data they have collected and the interpretations they have proposed. Preliminary versions of the manuscript were subjected, several years ago, to the penetrating critiques of Drs. Klaus Hoffmann and John Thorson. Later versions have benefitted greatly from comments of Mr. Curtis Baker, and Drs. G.D. Lange, S. Daan and J. Aschoff. Dr. E. Gwinner offered valuable advice on Chapter 15. The contributions of Drs. John and Ann Thorson to the final revision of the text, both in form of substance, have been more valuable to me than any few words here can describe. That revision was undertaken during a Fulbright fellowship, at the Institut für Zoophysiologie der Universität Innsbruck, under the sponsorship of Dr. W. Wieser. My wife, Roswitha, has also given her generous support, not only in preparing all the figures, but also in encouraging me throughout the long periods of analysis and writing. My thanks to all.

February 1979   J.T. Enright

Contents

Chapter 1
Introduction

The day-night environmental cycle is one of the most pervasive of the many factors which have shaped the evolutionary adaptations of both plants and animals. The vast majority of organisms are specialized in one way or another so that they can best carry on their primary life functions during only a fraction of that daily cycle. Plants, because of their dependence on sunlight for photosynthesis, are of course strictly diurnal in their basic physiology; among animals, some have chosen the option of being diurnal and others are nocturnal in their physiology and behavior. In order for such temporally specialized creatures to survive at minimal cost that portion of the day-night cycle to which they are less well adapted, essentially all have developed some form of the strategy, "Rest and wait until good times come again." In the majority of animals, this involves seeking shelter from the adverse environmental situation — including its predators — and reducing the intensity of metabolic activity, saving energy for tomorrow. In its most extensive development, among higher vertebrates, this resting phase of an animal's daily routine is what we know as sleep.

Sleep can, of course, be studied on its own, as a physiological state with many unique and interesting properties. The nature of sleep, its causes, concommitants and consequences, have indeed been of central concern to mankind since the earliest of times: the philosopher has speculated, as well as the poet, the singsmith, and the naturalist; and, in recent decades, the physiologist has investigated. In this scientific exploration of sleep, Nathaniel Kleitman was one of the pioneers, and his monumental book, *Sleep and Wakefulness* (Kleitman 1963), from which the title of this monograph was derived, represents a landmark in the field. There he synthesized a viewpoint based upon 40 years of his own research, and a bibliography of more than 4000 references. As indicated even in the book's title, Kleitman viewed sleep as only one facet of the broader phenomenon: the regular alternation between the sleeping and waking states. It was a central theme of his career that no satisfactory understanding of sleep can be achieved without a concommitant understanding of wakefulness.

It is self-evident, of course, that adequate definitions and characterizations of sleep must, at least implicitly, invoke a contrast with wakefulness, but the basic issue involved in Kleitman's viewpoint is more fundamental. In some of his earliest researches, Kleitman (1923) noted that during prolonged sleep deprivation the physical deterioration of the subject is not gradual and monotonic. Instead, a waxing and waning appears, which follows the succession of days and nights. The *tendency* to sleep or to be wakeful thus varies cyclically due to internal factors; sleep itself is not an essential component of that rhythmic variation, but only the *usual* manifestation of one phase in a cycle, a cycle which persists independently of sleep itself.

The physiological mechanisms which underlie this tendency to sleep or to be awake are a central concern of this book. The approach here, however, departs markedly from the usual focus of what is today known as "sleep research"; rather than attending to the details of what occurs during sleep itself, emphasis will be placed here on the *timing* of sleep and wakefulness, and the rhythmic alternation between the two states.

In its evolutionary origin, this rhythm is clearly an expression of adaptation to the day-night cycle: an endogenous oscillation in physiological properties of the organism, which can persist in the absence of environmental forcing – and which, in brief, is what is known today as a circadian rhythm. A number of researchers (see, particularly, Aschoff 1965a) have demonstrated that the human wake-sleep cycle will in fact persist as a regular rhythm for many weeks, when a subject is isolated under constant conditions and without external time cues. Such studies of human subjects, however, are relatively few in number, when compared with the extensive experiments which have examined the circadian rhythms of other higher animals. The majority of these animal studies have also dealt, if somewhat indirectly, with the wake-sleep cycle. Sleep has rarely been monitored directly, by measuring the animal's electroencephalogram (for exceptions, see Coindet et al. 1975; McNew et al. 1972; Mittler et al. 1977; Borbély and Neuhaus 1978a, b); instead, locomotor activity has been monitored, which provides an indirect index of when the animal is awake. Chapter 2 will elaborate on the methods used in collecting such data, and their implications about sleep and wakefulness.

It is worth noting at this point, however, that the terminology associated with such studies serves with remarkable effectiveness to obscure any relationship to other sorts of sleep research. The wake-sleep data from a biological-clock study are called an activity rhythm, with no explicit recognition of the rhythmic recurrence of sleep. The two components of a cycle are called rest time and activity time, with no direct reference to sleep and wakefulness. The results are published under the rubric of circadian rhythmicity, with no indication that the data might be of interest to those who care primarily about sleep. These kinds of experimental work which deal, in essence, with the *when* of sleep have thereby become deeply intertwined with other aspects of circadian rhythmicity, based on the urge to examine biological clocks in an evolutionary context. The spectrum of presumably related phenomena covers the range from the leaf-movement rhythms of plants to the eclosion rhythms of flies; it is thus not a simple matter for those interested in sleep to extricate the relevant aspects of circadian studies from the context of other phenomena which appear at first glance to have little to do with the mechanisms underlying sleep.

The consequence of these differences has been that those who have studied the detailed mechanisms of sleep itself, and those who have studied when sleep normally occurs, have tended to go their separate ways, each group no doubt convinced that it has little to learn from the other. In my opinion, the existence of this dichotomy in the field, as understandable as it may be, is artificial, unfortunate, and without rigorous logical justification.

This is a book about mechanisms which may underlie the *when* of sleep − and of wakefulness − and I hope, with this introduction and the ensuing chapter, to call to the attention of those who study sleep and those who study activity rhythms that their areas of interest are not as irrelevant to each other as the current literature might lead one to believe. Their potential mutual relevance derives from very elementary considerations. Modern sleep research, *sensu stricto,* accepts that sleep − and wakefulness − originate primarily from structures and processes of the central nervous system. Similarly, the activity rhythms of higher vertebrates are now widely believed by circadian researchers to be the result of a "pacemaker", also envisioned as a structure located in the central nervous system. The interrelatedness of activity and wakefulness makes it not only conceivable but − in my opinion − highly plausible that the mechanisms responsible for the two rhythmic phenomena − the wake-sleep cycle and the activity-rest cycle − have major common structural and functional elements at their core.

The empirical background for this exploration of the wake-sleep cycle is drawn exclusively (and necessarily) from the circadian-rhythm literature, and represents an array of data from many different laboratories. Experiments with animals, primarily birds and rodents rather than humans, provide the critical foundation for the present treatment because of the vast amounts of data available from long-term experiments with such subjects, derived in part from important kinds of manipulation not yet undertaken with human subjects.

Some readers may have misgivings about the extent to which any animal experiment is relevant to the physiology (and psychology) of human sleep. For the adult human, the most conspicuous − and most disconcerting − aspect of sleep is the loss of consciousness. The question of whether an animal's experience of sleep includes this feature cannot be determined until consciousness can be rigorously defined, in terms of all its physiological correlates. Nevertheless, to the layman, the sleep of birds and rodents appears quite comparable with human sleep: a lethargic state characterized

by great reduction in responsiveness to stimuli; furthermore, the wakefulness of a bird or rodent also seems qualitatively comparable with human
wakefulness.

In addition to these obvious superficial resemblances, the data of physiologists tend to support the layman's impression, that the sleep of all higher vertebrates is similar; important resemblances even exist in the major
features of the electroencephalogram. Although there are, of course, quantitative details that differ from one species to another, most sleep researchers today agree that birds and mammals do indeed sleep; comparative data
offer no clear justification for suspecting that the wake-sleep cycles of
birds and mammals are fundamentally different in nature from those of
man.

Based on experimental data on animal activity rhythms, the substance
of this book is an exploration of some of the consequences of one central
hypothesis, which is the following: that the wake-sleep cycle of higher vertebrates reflects the activity of a central-nervous-system pacemaker, which
is composed of an array of coupled neuronal oscillators with circadian period. More explicitly, I postulate that these are neurons which alternate
between a discharging and a quiescent state on a circadian basis, and which
interact through excitation so as to produce a mutually entrained system,
the output of which is primarily responsible for wakefulness and sleep.

While this central hypothesis can be briefly stated in such qualitative
terms, it must be reformulated in quantitative detail, in order to determine
the consequences of the postulate with any degree of specificity. This kind
of quantitative detail constitutes the structure of a model, defined in Chapter 4, in which explicit equations play a central role and by which consequences of the assumptions are explored through computer calculations in
subsequent chapters.

In contrast with several sorts of models which have been previously proposed for biological rhythms (e.g., Winfree 1967; Pavlidis 1969, 1971, 1973;
Wever 1962, 1963, 1964), the essence of the approach here involves "Monte
Carlo" simulation, in which stochastic (i.e., random) events play a major
role in the outcome. In an extended game of chance (hence the name,
"Monte Carlo"), the short-term outcome is intrinsically unpredictable; one
can only formulate expectations for the long-term average result. Similarly,
no two Monte Carlo simulations of the detailed behavior of a given system,
in which each event is simulated separately, are completely identical. The
variability in detailed outcome means that each single result can be considered the equivalent of a numerical experiment on the computer. The
search for generalizations requires the examination of an array of specific
cases; replication becomes an essential element in evaluating the validity of
conclusions, just as it is in any branch of experimental science.

While I have undertaken the detailed study of these hypothetical systems by means of numerical simulation on a computer, this has been a choice based on convenience rather than necessity. With a modest degree of ingenuity in circuit design and the appropriate electronic components, it should be a relatively simple matter to construct physical analogues of these models. Such an analogue would then permit electronic simulation, rather than simulation on a digital computer. The major disadvantage to that approach is that real-hardware experiments would require enormously more time.

The objective here will be to convey an empirical understanding of a novel kind of interactive network, the behavior of which is based upon familiar properties of nervous tissue. Only those components and behavior are invoked which seem plausibly derivable from what is already known by neurophysiologists, with a single critical exception: the assumption that ultra-low-frequency neuronal cycles (in the circadian range) exist and are qualitatively comparable, in a formal sense, with what is known of high-frequency spontaneously firing neurons. Granted that as a starting point, the construction elements for the postulated pacemaker, the nuts and bolts, will be drawn only from readily available materials listed in a catalog provided by neurobiology. Given such elements, the issue becomes one of how they might be interconnected to do the required job.

The most widely recognized criterion for the success or failure of an attempt to model a complex system is the extent to which new and unexpected predictions result, which may subsequently be confirmed or disproven by future experiments. The derivation of such predictions will not be slighted here (Chaps. 13 and 14), but predictive capacity is by no means the sole criterion that should be used to evaluate a model. One has the right to expect that the model first be able to account adequately for a significant body of existing data, without invoking additional assumptions for each additional experimental result. A primary question of interest, then, is the extent to which a large and complex set of observations can be regarded as natural consequences of elementary and plausible starting assumptions. Major portions of this book are devoted to this issue; it will be shown that the basic kind of model involved here can serve as a unitary framework for the interpretation of many diverse sorts of existing experimental data. The agreement between performance of the models and the behavior of the animals is in several ways both extensive and detailed. As a final use to which an otherwise successful model can be put, one can — once the necessary and sufficient properties of the components are determined — undertake a search for corresponding components in the animal. Suggestions a about how that search might be narrowed and guided by the results from computer simulations will also be offered, together with an exploration of one case in which a possible morphological counterpart of the models may have already been discovered (Chap. 15).

Chapter 2
A Description of Activity-rhythm Recordings and Their Implications

As indicated in the preceding chapter, extensive experimental data were published during the past twenty years, which deal with what are now known as the circadian activity rhythms of small birds and mammals. For the benefit of those not familiar with these kinds of experiments, a few words about methods and their limitations are essential.

The data which will initially be of interest involve "free-running" rhythms, obtained under noncyclic experimental conditions: the animal is kept in isolation, with constant temperature and constant light (or darkness), and with food and water continuously available in excess. Outside stimuli, such as noise which might disturb the animal, are kept to a minimum, and the interventions associated with renewal of food and water are as infrequent and nondisruptive as can be conveniently achieved. In other words, those obvious factors which might inform the animal about time of day, or otherwise perturb its "clockworks" are eliminated. Under such circumstances, the organism, whether human or monkey, rodent or bird, typically shows a well organized wake-sleep cycle, which consists of a several-hour stage during which gross motor activity is concentrated, followed by a sustained block of rest or sleep. This cycle repeats itself, with a period on the order of 24 h. Most such rhythms are "self-sustained", in the sense that they persist, without evidence of damping, for as long as the recording is continued.

The primary variable measured in such an experiment, and commonly the only one, is the animal's locomotor activity, and interest is usually focussed predominantly upon *when* the animal was active, *when* inactive, rather than how active. The method of monitoring activity varies with the species. For registration of the activity of birds, the holding cage is usually equipped with a perch supported by microswitches, which are sensitive enough that the weight of the bird on the perch closes an electric circuit. For experiments with mammals, the cage is commonly supplied with a running wheel, each revolution of which trips a switch. These electric on-off signals, from either perch or running wheel, are then monitored by an event recorder, through which a strip of chart paper slowly and continuously advances. A typical chart speed for the recorder is 3/4 in. per hour, meaning that an entire day's activity record is condensed into a strip of paper 1/4 in. wide and 18 in. long. In principle, data generated in this way could be presented as a single continuous record, with each hour's recording shown successively from left to right, exactly as it comes off the recorder, but data in that format, even from a few days, are difficult to assimilate at a glance, and inconvenient for publication. To deal with these limitations, a standardized method of data reduction has been developed, in which each day's record, with a time scale in hours, reading from left to right, is placed directly *beneath* the record from the preceding day. This kind of data presentation is sometimes referred to as an actogram; it is a compact condensation, even for very long records, and it offers other advantages as well. For example, if an animal were to awaken each day at the same exact clock hour, its awakening times would constitute a precisely vertical series of points in the actogram. Hence, a rough visual estimate of the average period of an activity rhythm can be made at a glance, from the day-to-day slope of the activity pattern, provided that the period is relatively near to 24 h. Actograms of this sort, both real experimental data and those resulting from computer simulation, are presented throughout this book, the first of them in Figures 2.1 and 2.2.

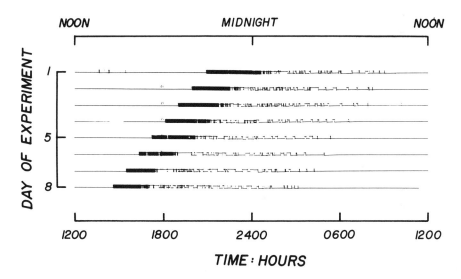

Figs. 2.1 and 2.2. Actograms showing the perch-hopping activity of house finches (*Carpodacus mexicanus*) under constant dim light. The lower position of the pen-line represents closure of the switch, meaning that the bird was on the perch. The wide, solid bars at the start of each day's activity represent such intense activity — several 100 contacts per h — that the individual pen deflections blur together

Simplicity is the most conspicuous virtue of this means of data collection. It has now become a routine matter to obtain records over many months from dozens of animals simultaneously. The capital outlay is small; and the necessary operating expenses consist only of food for the animals, recording

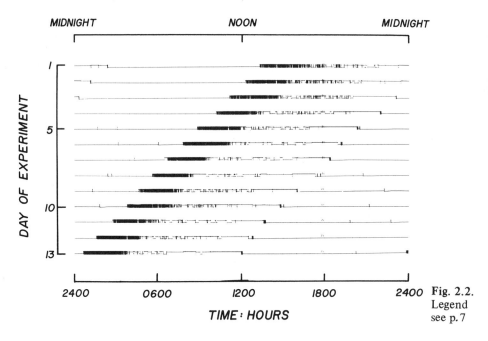

Fig. 2.2.
Legend
see p. 7

paper, and an hour or two a day of a technician's time to assemble the records (although some well-financed laboratories have interposed a computer to handle the steps between registration cage and assembled actogram). There are, however, limitations of the experimental procedure which significantly weaken interpretation of the data in terms of the wake-sleep cycle.

Clearly, when a bird is hopping onto and off a perch, or when a rodent is using its running wheel, the animal is awake, but lack of such activity does not necessarily indicate that the animal is asleep. The problem is not trivial; it has perhaps been a critical factor which has led those interested in sleep to reject the data as uninterpretable for their purposes. Obviously, if one were to rely on absence of locomotor activity as the only indicator of when a human subject is asleep, the data would border on the meaningless. The locomotor performances of birds and rodents, however, are usually somewhat less ambiguous.

In recordings from finches (Figs. 2.1 and 2.2), it is quite common for the bird to spend its entire "inactive" time sitting quietly on the recording perch (as indicated by the lowered position of the baseline in the recordings). It therefore becomes evident that the finch has remained essentially motionless for the entire inactive interval; it has not fed nor hopped aimlessly about the floor of the cage (where its inactivity would go unregistered); and it has rarely even fluttered sufficiently upon the perch, during this "rest time," to cause even single millisecond breaks in the electric circuit. At the least, the observed level of inactivity is sleep-*like*.

The bird may, of course, have been fully awake, sitting on the perch and surveying its modest and spatially limited domain, but if so, there is no overt indication of such perception, no effort to investigate or act upon anything seen. Furthermore, intermittent direct observation in some experiments of this sort has confirmed that prolonged inactivity is, indeed, usually a sign of sleep, with the bird having its head under its wing (Wahlström 1964). Since all forms of external stimulation remained completely constant, the sustained inactivity also suggests indirectly that the bird's responsiveness (and presumably its sensitivity) to stimuli were probably greatly reduced: an interpretation which has been confirmed by the finding that deliberate massive disturbances of a bird during such intervals of inactivity are responded to with extreme lethargy (Wahlström 1964). In brief, then, a bird's prolonged intervals without locomoter activity seem to fulfill those first-order criteria proposed by Piéron (1913) as a definition of sleep, and endorsed by Kleitman (1963): ". . .an almost complete absence of movement and an increase in thresholds of general sensitivity. . . ." These criteria would not alone be accepted as adequate in any modern laboratory devoted to sleep research, *sensu stricto;* they are necessary but not sufficient criteria for unequivocal recognition of sleep. Nevertheless, in the absence of long-term experiments in which sleep is more rigorously evaluated, by the more usual method of electroencephalogram (EEG) recording, locomotor-activity records from birds offer a useful basis for cautious inferences about the wake-sleep cycle.

Somewhat more restrictive qualifications must be attached to records of running-wheel activity like Figures 2.3 and 2.4, since the squirrel, while in its nesting quarters and not activating the wheel, may have undertaken a variety of waking motor activity without being asleep. Other recording procedures, which occasionally have been used to monitor both running-wheel activity and all movements in the nest box, demonstrate that a good deal of spatially-restricted activity does indeed take place outside the wheel. While some of that nest-box activity may be nothing more than the equivalent of a human's rolling over in his sleep, this is only a partial explanation, since the animal must leave its wheel to eat and drink. Hence, absence of wheel-running activity, in the final analysis, can do no more than suggest that the animal *may* have been asleep.

The conceptual problem in linking "activity-rhythm" data with the wake-sleep cycle on a long-term basis is not, however, as acute a difficulty as the preceding reservations might suggest, simply because wheel-running activity, as well as extensive perch-hopping activity, unequivocally demonstrate that the animal was awake. This definitional identity means that the wake-sleep rhythm and the activity rhythms are inescapably "coupled", and the period of one can be approximately evaluated from the other. For example, suppose a rodent, under constant conditions, begins its locomotor activity an

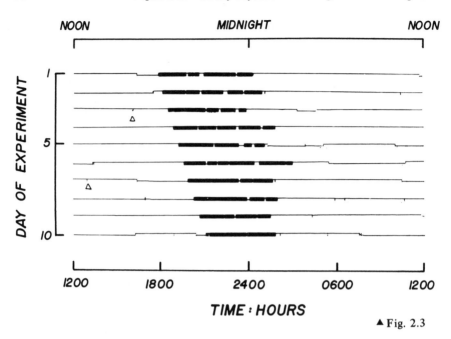

▲ Fig. 2.3

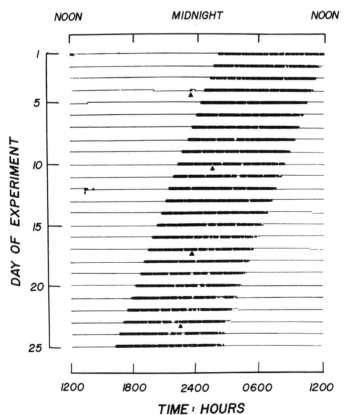

Fig. 2.4.

hour earlier on each successive day, over an interval of several weeks, and the period of its activity rhythm is therefore estimated as 23 hours: then it is a legitimate inference that the average period of the wake-sleep cycle was also approximately 23 hours. One can entertain the possibility of modest changes in the phase relationship between the two overt rhythms, on a cycle-to-cycle basis, such that the locomotor-activity time might "slip around" within the duration of wakefulness, but any persistent difference in period of the two rhythms would, in the long run, imply a lengthening of the duration of wakefulness, to the point that sleep must eventually vanish (Fig. 2.5); or require one to postulate that the animal was active when asleep.

It remains conceivable that two different timing centers exist within the organism, one governing the rhythm in locomotor activity and another responsible for the wake-sleep cycle, but if so, it is evident that they must be intimately interconnected: so closely linked that neither timing center can free-run independently for any extended interval. Furthermore, we know, from sleep-deprivation experiments like those of Kleitman (1923), that sleep itself is not a necessary component in the processes which ordinarily lead to the rhythm in sleep and wakefulness; it is only the *usual* manifestation of a set of conditions, which continue with their normal rhythmic timing even without sleep. Nor, as we shall see in the next chapter, is locomotor activity an essential link in the timing of activity rhythms; it, too, is a possible but nonessential indicator for the state of central timing processes. Bearing in mind that wakefulness is a necessary precondition for locomotor activity and that inactivity is a precondition for sleep, I find it far more economical to presume that there is a single rhythmic center responsible for the timing of both kinds of phenomena, than to think in terms of two intimately coupled centers, each independently rhythmic.

It would be misleading, however, to gloss over the large measure of caution which is necessary in interpreting recordings of locomotor activity as indicators for the wake-sleep cycle. While long-term activity records permit unequivocal estimates of the average period of the wake-sleep cycle, the exact timing and duration of sleep, in any cycle, cannot be inferred from activity records; and, of course, no information is provided at all about the depth of sleep and its stages.

There is, therefore, no doubt that if an equivalent body of data existed, in which sleep and wakefulness had been continuously evaluated by more

Figs. 2.3 and 2.4. Actograms showing the running-wheel activity of flying squirrels (*Glaucomys volans*) in complete darkness. The solid horizontal bars represent multiple pen deflections which occur so frequently that they blur together. *Closed triangles* represent disturbances without light; *open triangles* represent disturbances with dim light. Illustrations based on DeCoursey 1961a and DeCoursey 1961b respectively

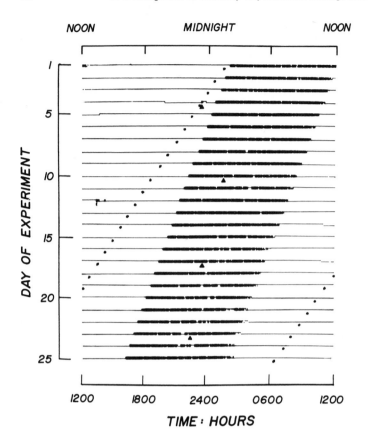

Fig. 2.5. The actogram of Figure 2.4, with a superimposed series of dots to indicate hypothetical times of awakening, if the period of the animal's wake-sleep rhythm were to be 23 1/4 h: about 20 min shorter than the period of the activity rhythm. The actual duration of wakefulness would extend from these points to the end of daily locomotor activity, if not farther; there would be an ever-increasing gap of time between awakening and onset of wheel running, until eventually awakening would occur so as to coincide with end of the preceding bout of wheel-running activity. Note that if the period of the wake-sleep rhythm were assumed to be *greater* than that of the activity cycle, the animal would eventually be required to begin wheel running while still "asleep"

rigorous procedures, those would be the data of preference as the empirical basis for the present treatment. Such data are not, however, available, no doubt in part because of the technical problems both in obtaining very-long-term continuous EEG records, and in evaluating such records, once obtained. Given the absence of comparable, long-term EEG data for the direct measurement of sleep-wake cycles, it seems evident that there will, at the least, be strong correlations between activity cycles and the presently unavailable sleep-cycle data. The objective in subsequent chapters will, therefore, be to explore a class of models which can account for important aspects of activity-rhythm data, based on the conviction that one is thereby, *at least approximately,* accounting for the wake-sleep cycle as well. Perhaps the treatment here will even provide an incentive for the collection of comparable, long-term data on sleep and wakefulness, more rigorously measured, so as to determine which important temporal features of those data — if any — the present kinds of model cannot account for.

Chapter 3
The Pacemaker and its Precision

The central hypothesis of this book invokes a "pacemaker" as the source of the wake-sleep cycle. It is a matter of definition that a pacemaker is a persistent generator of regular, rhythmic output, and so at first glance the proposition that the wake-sleep cycle of vertebrates reflects a pacemaker may seem to convey nothing more than that a rhythm of some regularity has been observed. There is, however, a critical additional implication, testable by experiment, which is that wakefulness and sleep, activity and inactivity, are not themselves essential elements in the "feedback loop" which is responsible for the observed rhythm. Applied to the human experience, it is an initially appealing hypothesis that the duration of sleep is determined only by how tired one was from the preceding day's activity; that the duration of wakefulness is then determined by how long and soundly one has slept, and how much activity has been undertaken before being again tired enough to sleep; and that the regular sequence of these two independently timed processes might lead to the daily wake-sleep cycle. This suggestion *seems* to comport with our ordinary experience, but it is exactly this kind of proposition that the pacemaker concept is intended to contradict. Instead, the term pacemaker implies that there is within the organism some rhythmic center, the output of which ordinarily is directly associated with sleep and wakefulness; but with the further specification that this structure is very little influenced, if at all, by whether the organism *in fact* responds by wakefulness or sleep, activity or inactivity. Extreme exertion may, of course, lead to premature onset of sleep in any single cycle; unusual external stimuli may lead to premature awakening, perhaps even in the middle of the night; but the pacemaker proposition implies that these unusual extraneous events do not reset that center which ordinarily "drives" sleep and wakefulness. Subsequent cycles would then be little affected in their timing by superimposed interruptions of the driven overt behavior.

Evidence for the existence of such a pacemaker for the wake-sleep cycle was recognized very early by Kleitman (1923), as already mentioned in Chap. 1: during sleep-deprivation experiments, a cyclic circadian variation in degree of arousal, in the urge to sleep, persists even in the absence of sleep itself. Evidence of another sort is available from experiments with drugs, for which examples from the canary are presented in Figure 3.1: in spite of the fact that barbital clearly delayed awakening on the first day

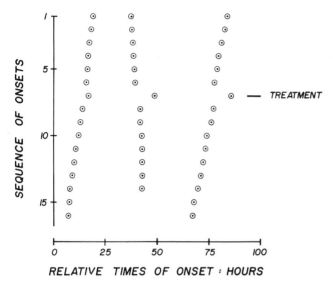

Fig. 3.1. Three examples of the effect of a single administration of the drug barbital on the timing of activity onset in canaries. As in an actogram presentation, successive onsets of locomotor activity are plotted beneath each other on the basis of a 24-h cycle; the abscissa values for the starting points on day 1, however, are arbitrary. The abscissa is just a scale, modulo 24 h, to permit comparisons of timing within a sequence. Note that a delay of as much as 6 h occurred on the day of treatment, but on the following day the timing of the animals' rhythms shows little if any residual effect; the pacemaker is *not* reset by the delay in awakening. Data from Wahlström 1964

following its administration to these birds, the data indicate unequivocally that some rhythmic center, which ordinarily controls awakening, was largely unaffected by this delay.

A generator of regular rhythms which is self-contained, and which therefore does not depend critically in its timing upon those peripheral events which it evokes, is referred to by physiologists as a pacemaker, and it is the properties of this pacemaker which are of interest here. Overt manifestations such as sleep and wakefulness, activity and inactivity, can do no more than provide clues about the status and operation of the pacemaker. As Figure 3.1 clearly implies, these overt indicators can under some circumstances be potentially misleading, but since the pacemaker has not yet been directly observed, we usually have nothing more than such clues to rely upon.

While evidence like that in this figure demonstrates that vertebrate activity rhythms are a reflection of some kind of pacemaker, the nature and location of that pacemaker are open questions. The assumption will be made here that there exists a specialized structure located in the central nervous system which serves as the pacemaker, but that assumption is by no means dictated by unequivocal experimental evidence. There are, of course, many indications that certain structures in the brain of higher vertebrates are involved in some way in sleep and wakefulness, as well as in the persistence of free-running rhythms of locomotor activity, but there is no basis for unanimity among experimentalists on the location or identity of any single principal structure which can be equated with *the* pacemaker of interest here. It is not even certain that a single, discrete structure is responsible for pacemaker function; there might, for example, be two semiautonomous and mutually inhibitory centers, one responsible for sleep and the other for wakefulness; or blood-borne hormones might be essential links in the circuitry of the pacemaker. There is, in fact, clear evidence that hormones can at least modulate the performance of the pacemaker, producing modest changes in its period (Turek et al. 1976) or its pattern of output (Gwinner 1974); and some data from birds even suggest that circadian variation in the level of a hormone may be a critical feature in the functioning of the pacemaker itself (Zimmerman and Menaker 1975). This latter case and its possible relationship to the models to be developed here will be subsequently treated in detail (Chap. 15); for the time being, I simply assume that one pacemaker system may be sufficient to account for the data of interest, and assume further that this system is (or at least can be treated as though it were) fully contained in structures of the nervous system.

At this point, we turn from the realm of assumption to a central and most remarkable empirical fact. Circadian activity rhythms often show an extreme precision over many cycles, as documented in Figures 2.1 to 2.4. This is not to say that they reliably measure off 24 h in each cycle. As the term "circadian" indicates, these internally generated rhythms usually do not have a period of *exactly* 24 h; instead, their free-running periods often differ by an hour or more from 24 h. The cycle-to-cycle *precision,* however, the repeatability of period length in successive cycles, is often astonishing, when compared with that of other, more familiar rhythmic phenomena in biologic systems. For example, in the records of both Figures 2.1 and 2.3, the standard deviation of period length, measured on the basis of activity onsets, is about 3 min; and only slightly greater variability is evident in Figures 2.2 and 2.4 (standard deviations of 7 and 4 min, respectively): values which represent an "error" of 1 part in 200 to as little as 1 part in 500.

Those examples have been selected as unusually precise performances. Comparable precision has not yet been observed in the activity patterns of invertebrates or lower vertebrates; and even among birds and mammals, there are major differences among species, as well as among individuals of the same species, in the day-to-day precision of free-running rhythms. Even granting, however, that the animals which produced the recordings shown in Chap. 2 are somewhat unusual, the observation that such temporal precision is sometimes achieved by organisms is most remarkable. Modern engineering can, of course, do much better, but most biological systems do not,

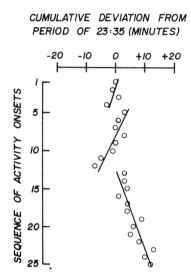

CUMULATIVE DEVIATION FROM
PERIOD OF 23:35 (MINUTES)

Fig. 3.2. Successive onsets of locomotor activity from the recording of Figure 2.4, replotted as deviations from an average period of 23 h 35 min. The two discontinuities in this enlarged presentation are correlated with disturbances of the animal on days 4 and 12. Detailed data kindly provided by Dr. P. DeCoursey

no matter what the time scale. A more detailed examination of the data in Figure 2.4 demonstrates that overall calculations of precision based on such records sometimes may actually even underestimate the precision of which the pacemaker is capable. In Figure 3.2, the expanded plot of successive times of onset suggests that the overall record might better be subdivided into three sections, each of appreciably greater precision than the overall single estimate. Two slight shifts in the rhythm seem to have occured, on days 4 and 12, each associated with a disturbance during the animal's sleep, time (see symbols in Fig. 2.4). This proposed subdivision of the record is of course somewhat subjective, but in any case, a single trend line through the data would seriously oversimplify the complexity of pattern; a single, pooled estimate of average period will therefore lead to underestimates of the pacemaker precision. The interpretation, that the true precision of the circadian pacemaker may be even greater than the raw data imply, has also been supported by analysis of a large number of such records from other species. In most of those records, the sequential times of activity onset showed negative first-order serial correlation, a finding which can be interpreted to mean that an appreciable fraction of the variance in observed cycle length is associated with the indicator process (activity) and is not due only to the pacemaker (Pittendrigh and Daan 1976a).

The kind of reproducibility of period, documented in Figures 2.1 to 2.4, stands in stark contrast to the behavior of most other familiar biological rhythms. The property of temporal precision has, for example, been extensively investigated in the discharge patterns of single nerve cells. Neurons which fire spontaneously under constant conditions commonly have a maximum precision poorer than one part in ten (ratio of standard deviation of

inter-spike intervals, s, to mean interval, I; see Enright 1967, and the literature cited there). Typical of these single-neuron data is a tendency for a
given cell to be most precise (minimum s/I) when it is firing at highest frequency; as the level of tonic stimulation decreases, thereby increasing mean
interval between spikes, precision decreases. Variability of less than 1 part
in 50 is almost never encountered in such single-cell data, and if seen, is the
basis for classifying the neuron as an exceptionally precise pacemaker. Circadian activity rhythms can therefore be about ten times as precise as the
best of single-neuron recordings.

The cycle-to-cycle unpredictability in the performance of nerve cells has
led to several different models (ibid.), all of which include stochastic elements (i.e., random with respect to time) in the mechanisms underlying
spike generation. Those models are one mode of expressing the growing
conviction of many neurobiologists that an adequate understanding of nervous systems will require explicit inclusion of random processes in the dynamics. At a mechanistic level, the data leading to such models demonstrate
that even on the time scale of a few milliseconds to a few seconds, fluctuations in apparently uncontrollable properties of the cell or its environment
are of major importance to the dynamics of spike initiation. Furthermore,
essentially all single-neuron recordings, in vivo or in vitro, are characterized
by unpredictable fluctuations which can be called "baseline drift"; the
mean interval, which may be relatively reproducible over a few minutes,
seldom remains constant over longer periods of observation, in spite of all
attempts to maintain constant conditions. Such properties of single-cell
rhythms serve further to emphasize the remarkable stability of circadian
activity rhythms, which show great precision on a time scale of many days.
Somehow the intuitively expected sources of uncontrollable long-term variation, which might lead to major imprecision, appear to have been eliminated or compensated with remarkable effectiveness.

Imprecision of the sort seen in spontaneously firing neurons is also characteristic of a variety of other relatively regular biological rhythms which involve whole organ systems. For example, cycle-to-cycle variability of human
heartbeat in a sleeping subject is typically on the order of 1 part in 25 (ratio
of standard deviation to mean interval), with optimal precision obtained
in some short-term estimates being on the order of 1 part in 60 (Buck and
Buck 1968). Several examples of the heartbeat rhythm are shown in Figure
3.3, in a format comparable with an actogram, so as to facilitate visual comparisons with the precision of circadian actograms like those in Figures 2.1
to 2.4. The chirping of undisturbed crickets is usually no more regular, on
a cycle-to-cycle basis, than about 1 part in 30, and a similar value has been
obtained for repeatability of interval length in the persistent calling of the
whippoorwill (ibid.). The cycle-to-cycle variability in the human menstrual
rhythm is usually so great that it scarcely deserves mention in the present

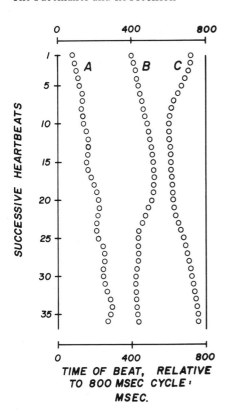

Fig. 3.3. Three examples of the heartbeat rhythm in resting human subjects. Cycle-to-cycle variability in the overall records (standard deviation/mean interval) is about 1 part in 20 for subjects A and B, and 1 part in 10 for subject C. Much of that variation, however, is contributed by low-frequency noise: gradual changes in mean interval. In the first 15 cycles for subject B, the variation was only 1 part in 75; and cycles 10 to 22 for subject C have a variability of 1 part in 50

context. In discussing the rhythmic flashing of certain fireflies, emphasis has been placed on the exceptional precision of the system (ibid.); the cycle-to-cycle variability can be as little as 1 part in 200. This is, however, precision measured over only a few minutes, and the best of these short-term performances observed of fireflies is no better than the poorest of the circadian records illustrated in Figures 2.1 to 2.4.

There is, as far as I have been able to determine, only one biological rhythm in which temporal precision greater than that of circadian systems has been recorded: the high-frequency discharge pattern of certain electric fish (typically several 100 cycles per s), where short-term reproducibility of cycle length can achieve the astonishing value of about 1 part in 8000 (Bullock et al. 1972). One cannot gloss over the precision of electric-fish rhythms with a casual statement about invariance of the "absolute refractory period" of nerve cells which generate the rhythm; even if true, such an observation is no explanation at all, since an absolute refractory period is presumably a characteristic of all single-neuron rhythms.

A full understanding of the factors responsible for the precision of electric-fish discharge pattern has not yet been achieved, but a good deal more is known than in the circadian case. The pacemaker which generates the fish's rhythm is located in the central nervous system, and consists of an array of many dozens of similar cells. The synchrony of their discharge, which generates the externally measured electric field, is achieved by "coupling" within the ensemble. The performance of a single cell of this pacemaker has not yet been measured in functional isolation from the ensemble (and it may be that they cannot oscillate in this mode), but the nature of the coupling suggests that the average intrinsic periods of such cells, if they can oscillate alone, would probably differ appreciably from each other. Furthermore, on the basis of other single-neuron recordings, there is reason to suspect that the cycle-to-cycle precision of an isolated single cell of the pacemaker might well be appreciably poorer than that of the entire ensemble. The existence of coupling within the pacemaker implies a kind of interaction which could serve not only to synchronize an array of disparate single-cell rhythms to a single common average value of period, but also, in principle, to produce system output appreciably more precise than that of any constituent cell. Qualitatively speaking, the idea is that a sort of "averaging" may be involved in the observed precision, since averages are less variable than their constituent values.

The suggestion that the form of interaction, known as coupling within a population of oscillators, might be involved in producing precise output from a biological system was apparently first made by Wiener (1958) in the context of regular waves in an EEG. Barlow (1961) and Winfree (1967) have invoked the idea as a qualitative explanation for the precision of circadian rhythms, without further elaboration. On careful examination, the concept of coupling proves to be an amorphous entity, meaning different things to different people; a statistician, for example, would probably define it in negative terms, as a lack of complete independence. In the formulation of certain previous models to describe systems of coupled oscillators, the effect and strength of the coupling have been represented by the coefficient of one term in an equation (e.g., Pavlidis 1973). This sort of definition leaves the exact nature of the process in very abstract form, equally applicable to oscillators which are electric, mechanical, chemical or physiological. One must specify how the coefficient might "map" onto physiological variables before the definition can be useful to the experimental physiologist.

My initial objective here will be to do just that: to reformulate the general hypothesis, that coupling among oscillators might underlie the precision of circadian rhythms, based upon postulated structural and functional details which are compatible with known properties of nervous tissue. In other words, the procedure will be to translate the mathematician's definition of coupling into the dynamics of plausible physiological substrate. Were that "translation" to be the only substance of this book, a natural response might well be, "So what else is new? " Many other, equally specific translations are possible, since the term coupling is so flexible. We shall see, however, that one particular mode of coupling follows quite naturally from reasonable physiological assumptions, that this mode could easily account for precision, and, most important, that it has a variety of other interesting and nonobvious consequences. A system based on such coupling can, with minor additional specification, account in detail for a large variety of well-documented properties of the circadian wake-sleep cycle of higher vertebrates, many of which have hitherto either remained unexplained, or have been treated only by an array of ad hoc hypotheses.

A detailed consideration of these broader and more interesting issues must be postponed, since it will require a thorough understanding of the particular system envisioned, and the manner in which that system can attain precision in its overall output. Nevertheless, in order to sustain the reader's interest through the next several, rather demanding chapters, which focus on details of structure and basic functioning of the models, I offer here an abbreviated description of some of the biological data which will be confronted, as a foretaste of things to come.

The following relevant phenomena have been observed under free-running conditions:

1. "Aschoff's rule": For diurnal animals, the free-running period of a circadian activity rhythm usually decreases with increasing light intensity; the brighter the constant light, the faster the animal's "clock" runs. For nocturnal animals, the converse is usually observed: the free-running period of the rhythm increases with increasing light intensity.

2. The "circadian rule": For diurnal animals, brighter constant light prolongs daily wakefulness under free-running conditions, and also increases the level of arousal, as indicated by the intensity of locomotor activity. For nocturnal animals, the converse is true: brighter light usually shortens the duration of wakefulness and decreases the intensity of activity.

3. Cycle-to-cycle precision: a) Onsets of locomotor activity (awakenings) are nearly always more predictable (lower cycle-to-cycle temporal variability) than ends of activity, under free-running conditions. b) The ratio of variability in onsets of activity to variability in ends of activity has been observed to decrease when period of the rhythm is shortened. c) Maximal precision of activity onsets has been observed at intermediate values of period of the rhythm.

4. Effects of bright light: Circadian rhythmicity of birds degenerates completely when they are kept in continuous bright light; the wake-sleep rhythm can persist indefinitely, but only under a restricted range of light intensities.

5. Lability of pacemaker period: a) Long-term drifts in the free-running period of an activity rhythm are often observed, as well as occasional long-term fluctuations in period. b) Slight but predictable long-persistent modifications in free-running period of an activity rhythm can be induced by

specific treatments. Both prior constant light intensity, and the period of a prior entrainment regime can result in such "after-effects."

The following phenomena have been observed under cyclic lighting regimes:

1. The activity rhythms of both diurnal and nocturnal organisms can be synchronized by a simple on-and-off lighting cycle, but sudden transitions of intensity are not necessary for entrainment. A small-amplitude, continuously varying lighting cycle can also synchronize a circadian activity rhythm.

2. An activity rhythm can be synchronized to whole multiples of the period of an imposed lighting cycle; that is, a period of, say, 24 h can be induced by a light cycle with a 12-h or an 8-h period.

3. When subjected to a very-low-amplitude lighting cycle, an activity rhythm often shows a complex form of incomplete synchronization; the activity rhythm "scans" the light cycle with a cyclically varying period, showing what might be termed frequency modulation.

4. There are conspicuous and consistent qualitative differences between diurnal and nocturnal animals in synchronized phase relationships between an activity rhythm and a light-dark cycle, as well as striking differences in the manner in which re-entrainment is achieved following a shift of the light cycle.

When an organism, under free-running conditions, is subjected to a single brief pulse of bright light, changes can be induced in the subsequent timing of the activity rhythm. The following phenomena have been observed in such experiments:

1. The time at which the stimulus is given (i.e., when, during the animal's sleep-wake cycle) determines whether any response occurs, and if so, whether it is an advance or a delay.

2. In general, nocturnal animals tend to show much smaller responses to such single-pulse stimuli than diurnal animals.

3. For nocturnal animals, a light stimulus at or slightly before the time of awakening produces a delay in the animal's pacemaker; for diurnal animals, the result is an advance.

4. Maximal shifts in the rhythm, both advances and delays, occur during the middle of the sleep phase for diurnal birds. For nocturnal rodents, essentially no phase shift at all is induced by light stimuli during the sleep phase; they respond only during the waking phase.

5. A single, unrepeated light pulse which *advances* the clock of a nocturnal animal requires several days to exert its full effect, but the effect of *delaying* stimuli are completely expressed within a single circadian cycle.

These, then, are examples of the data with which we will eventually be dealing. It will be shown in subsequent chapters that models of the sort considered here can readily account for such experimental results, and others as well.

Chapter 4
A Class of Models for Mutual Entrainment of an Ensemble of Neurons

In the preceding chapter, the question has been raised: How can the circadian wake-sleep rhythm be so precise in its timing? A possible answer to this question, to be considered here, is that one might be able to construct a very reliable pacemaker out of an array of cellular components which are individually far less reliable: units which are very erratic in their cycle-to-cycle behavior, and which, while similar to each other in basic functioning, differ appreciably one from another in their quantitative characteristics. If a precise pacemaker is to emerge from the behavior of a heterogeneous ensemble of "sloppy" cellular elements, those components must interact with each other in some sort of a nonadditive fashion, and this chapter is ultimately concerned with a formulation of how coupling of that sort might take place. As a necessary prelude to that formulation, however, we must consider in some detail the behavior of the erratic elements of which the ensemble is composed, when they are acting in complete independence of each other.

These individual elements of the ensemble will be referred to as "pacers", and it is my assumption that each pacer is a neuron which is capable of independent oscillations with more or less circadian period. It must be admitted at the outset that there is, to date, no unequivocal empirical demonstration that *any* single neuron has the endogenous capacity to generate circadian rhythmicity, without cyclic interactions with its neighbors or milieu. Thus, this proposition involves a far-reaching extrapolation beyond the limits of experimental neurophysiology; nevertheless, certain lines of indirect evidence can be mustered which bear upon the proposed assumption, and rescue it from the realm of the highly implausible.

The most valuable of this evidence lies in the well documented finding that single cells, which are *not* neurons, can independently generate circadian cycles. A variety of unicellular organisms, predominantly algae, show sustained circadian rhythms under constant conditions, both as populations and as isolated individual cells (e.g., Karakashian and Schweiger 1976). Hence, multicellular interaction is not an inescapable prerequisite for the long time constants underlying circadian rhythms.

There is, furthermore, no doubt that nerve cells within an intact bird or mammal show clearcut circadian periodicity. Because of the multifaceted consequences of the wake-sleep cycle for both sensory and motor systems, one can infer that there must be vast numbers of cells in the central nervous system — perhaps even the majority — which change their behavior on a circadian basis. Most of the circadian neuronal rhythms that could be measured in the intact animal, however, would certainly be consequences rather than causes of the whole-animal circadian rhythm. Such rhythmicity in the nervous system does not, then, bear directly on whether single neurons, acting *independently,* can generate cirdadian cycles, as assumed here.

In the final analysis, the only fully adequate direct evidence justifying the assumption of interest would involve complete isolation of a single nerve cell, followed by the demonstration of circadian variation in output over several days. A small step in the direction of such a demonstration has been taken in the well-publicized case of that giant neuron, known as R-15, in the abdominal ganglion of the sea slug, *Aplysia.* Originally it was claimed (Strumwasser 1965) that there is a persistent circadian rhythm in these sustained bursts of activity, when the ganglion is maintained in isolation, but recent research has raised serious doubts about whether a true rhythm exists in that preparation (Strumwasser, pers. comm.). Regardless of how that issue is ultimately resolved, the evidence is quite clear that this single nerve cell can generate prolonged bouts of discharge activity, alternating with many-hour intervals of quiescence (Strumwasser 1965; Lickey 1966, 1969; Lickey et al. 1971; Beiswanger and Jacklet 1975; Audesirk and Strumwasser 1975); that the timing of these bursts is determined by events which precede the activity by many hours (Beiswanger and Jacklet 1975; Audesirk and Strumwasser 1975); and that the timing of the bouts of discharge does not depend upon triggering by synaptic input from other cells in the ganglion (Strumwasser 1965). With a preparation of this sort, it remains conceivable that some subtler sort of interaction with neighboring cells is necessary for the observed bursts of discharge, but the evidence is at least entirely consistent with the interpretation that the recorded behavior of R-15 is an endogenous product of the single neuron, acting independently.

The indirect evidence of interest here can be summarized as follows: single cells which are *not* neurons are capable of independent circadian oscillations; nerve cells of an intact vertebrate show circadian rhythms, most of which are probably driven rather than independent oscillations; and the neuron R-15, *probably* acting independently, shows long-term fluctuations in its behavior with at least some of the properties to be expected of a circadian neuron. None of this evidence constitutes unequivocal support for the proposition that individual cellular elements of the vertebrate wake-sleep pacemaker would have the capacity, when acting in complete independence, to discharge in a more or less circadian pattern. Hence, the models to be described here are based only on the assumption that this is so.

The next issue which arises in seeking general congruence between neurophysiological data and a pacemaker model is how one should deal with the proposed erratic behavior of the pacers. In order to examine this question, it is useful to review here a broad class of elementary models which have been previously proposed to account for the stochastic behavior of sponta-neously firing high-frequency neurons. Models of this class all describe a "relaxation oscillator" which has no cycle-to-cycle memory of its preceding behavior: after each nerve spike, the system is presumed to return to an identical "starting state". From there, a recovery process is assumed to be-gin, by which sensitivity gradually increases.

As indicated in Figure 4.1, there are significant differences, between mo-dels of this general sort, in the description of the time course of the reco-very process (Hagiwara 1954; Calvin and Stevens 1965; Buller 1965; Geis-ler and Goldberg 1966; Viernstein and Grossman 1961; Weiss 1964; Ver-veen and Derksen 1965), but for purposes of the present discussion the state of this process can simply be represented by the variable R(t), which decreases monotonically with time since the last preceding nerve spike. A second component of these models for high-frequency neurons is the as-sumption that there is some constant tonic level of extrinsically generated excitation, here designated E, which is independent of the recovery process. Finally, the models presume the existence of a very-high-frequency "noise" process, here designated n(t), which constitutes random, additive, moment-to-moment variation in excitatory state. This noise is usually taken to have a Gaussian ("normal") distribution;[1] to be free from serial correlation; and to be due to local, submicroscopic events in the nerve membrane, which are globally unpredictable. A nerve spike is then presumed to be triggered as as soon as

$$E + n(t) \geqslant R(t),$$

at which time the sensitivity function, R(t), returns to its starting state and the spike-generating process begins over again.

Models of this general sort have been independently proposed by many researchers, and have proven valuable for describing several interesting and rather complex aspects of the behavior of high-frequency spontaneously firing neurons under various levels of tonic stimulation (Enright 1967).

---

1 The assumption of a Gaussian distribution of noise can be greatly relaxed without producing major change in the predictions of the models. The basic requirement is only a continuous probability-density function in which extreme values are less likely than those near the mean. Nevertheless, the Gaussian distribution has much to recom-mend it for models of these sorts, because of the statistical principle known as the central-limit theorem: as sample size increases, the means of samples drawn from any population, no matter what its distribution, will converge on the Gaussian distribu-tion. Hence, if a particular sort of macroscopic event is due to the sum of the effects of a large number of microscopic processes of variable magnitude, the macroscopic events should have magnitudes which can be approximated by the Gaussian distribu-tion.

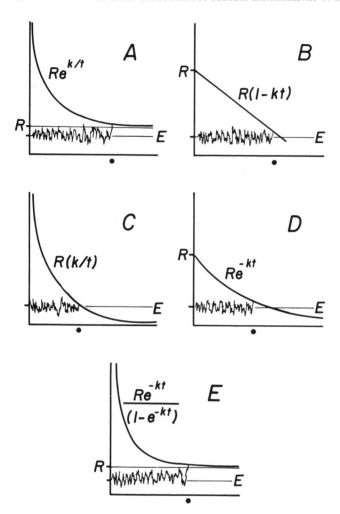

Fig. 4.1 A—E. Five related formal models which have been proposed to account for the temporal pattern of individual nerve spikes from neurons which show spontaneous activity and a discharge rate which increases with the intensity of tonic stimuli. The prevailing level of tonic stimulation is envisioned as determining average excitatory state, E, upon which random, high-frequency Gaussian noise is superimposed. Following each nerve spike, a monotonic recovery process (R) is initiated, which, in a loose sense, could be compared with threshold of the cell; the time-course of that process could presumably be assessed, in principle, by the supplementary imposed depolarization necessary to trigger another spike immediately. In the absence of such phasic stimuli, the occurrence of the next spike (*black dots*) is determined by the time at which tonic excitation (plus noise) exceeds the instantaneous value of the recovery process. These five models, proposed by Hagiwara (1954): Scheme A; Buller (1965): Scheme B; Calvin and Stevens (1965): Scheme C; Viernstein and Grossman (1961), Weiss (1964), and Verveen and Derksen (1965): all Scheme D; and Geisler and Goldberg (1966): Scheme E, differ only in the formulation of the time-course of the recovery process. Any of the models (except that shown in Scheme B) can, with appropriate choices of parameter values, adequately describe large bodies of rather complex experimental data from high-frequency spontaneous neurons, although for certain data sets the formulation of Scheme D proves most satisfactory (Enright 1967)

Should excitatory state (E) increase, due to greater tonic stimulatory input, the mean interval between spikes will, of course, decrease (i.e., frequency of firing will increase), and there may well be, in addition, changes in the variability of inter-spike intervals. The magnitude of n(t), expressed, for example, as the variance of the Gaussian distribution, is a prime determinant of how variable the intervals between spikes are. If the recovery process is relatively slow, and n(t) not too large, then the predicted distribution of intervals between successive nerve spikes may be approximately Gaussian. If n(t) were to become vanishingly small, the model would no longer be stochastic; inter-spike interval would be determined only through attainment of the prevailing excitatory state by the gradually decreasing recovery process.

These formulations of course ignore many of the well studied details of the ionic and electric phenomena which actually precede and follow a nerve spike; their virtue lies in the demonstration that much of that complexity can be ignored, if one only wishes to account for the temporal patterns in nerve discharge. Because such models have proven so useful as simple and general ways of formally describing details of the stochastic performance of spontaneously firing neurons, the pacemaker models to be developed here rely upon closely related formalism. This use of models derived from high-frequency neurons is not associated with any presumption that the detailed mechanisms underlying the several-hour bursts of activity, which are generated by the circadian neurons hypothesized here, are necessarily identical to those associated with single spikes of high-frequency nerve cells. The details of mechanism, whatever their nature, are simply subsumed in an elementary formalism, much as they are in the models of Figure 4.1. Even the differences among the several models illustrated in Figure 4.1 will be ignored; the time course of the recovery process, which differs among the models, will not enter the calculations other than in the stipulation that it be a function which decreases monotonically with time since last discharge. The other essential features of those models which are to be invoked here are: (1) that a steady-state stochastic noise process modulates the time at which an event occurs, and can lead to a Gaussian distribution of inter-event intervals; and (2) that tonic stimuli do not affect the rate of the recovery process, but only the excitatory level, E, which must be attained by the recovery process before an event occurs.

Now in the models to be developed here, it is assumed that a pacer neuron can oscillate with circadian period; hence, a pacer must show at least two measurably different kinds of behavior in different parts of its cycle. From an operational viewpoint, these two portions of each pacer cycle might simply be designated as the "on" and the "off" phases, but in keeping with the interpretation that the pacers are neurons, the "on" phase of the circadian cycle of a pacer is presumed to be characterized by sustained nerve discharge. Hence, the "on" or active phase of a pacer cycle will be termed the discharging phase and the inactive, "off" portion of the cycle will be called the recovery phase, during which some critical accumulatory or synthetic process presumably occurs. As discussed subsequently (Chap. 8), one might postulate that the discharge phase is vanishingly brief, relative to the recovery phase; a circadian neuron might give a single large spike once a day. That limiting case, however, will not permit the kind of coupling interaction that is to be considered below, by which system precision can be greatly enhanced; nor does it lend itself naturally to the expected behavior of a pacemaker which drives the wake-sleep cycle, since neither sleep nor wakefulness is a vanishingly brief portion of the overt rhythm. Both the recovery and the discharging phases will therefore be presumed to be appreciable parts of a single cycle of the pacer. We will now proceed to specify in detail the expected behavior of a pacer in what will be called its "basic" state, that is, when acting in complete independence of all other elements in the ensemble.

Let us first consider the duration of the recovery phase, which will be here taken to be the analog of the inter-spike interval of high-frequency neurons. Because of the stipulation that the pacers are erratic, no two cycles of a pacer can be expected to be absolutely identical in duration of the recovery phase. If one were to examine the behavior of a single pacer over a large number of cycles, however, there will be a mean duration of the recovery phase, $\overline{X}$, and a cycle-to-cycle variability in its duration. A simple method for summarizing erratic and unpredictable variation in the duration of the recovery phase of a single pacer of this sort can be obtained by invoking the Gaussian (or "normal") distribution as follows:

$$X_{b,i,j} = \overline{X}_{b,i} + \alpha z_j \tag{1}$$

where $X_{b,i,j}$ is the duration of the recovery phase, in the independent or basic state (the subscript b), of the i th pacer in the j th cycle; $\overline{X}_{b,i}$ is the mean duration of the recovery phase, during the basic state, of the i th component; z_j is a normally distributed random variate with mean of zero and unit variance, with a value assigned anew to a pacer in each cycle; and α is the coefficient (standard deviation) which converts the unit-normal variate into cycle-to-cycle standard deviation in hours. Note that z_j reflects the noise process in the system, which has the consequence that the duration of the recovery phase of a pacer cannot be predicted exactly for any single cycle. As is assumed in models for high-frequency neurons, the implicit assumption here is that an array of small-scale events at some unobserved level of organization produces temporal variability at a macroscopic level, variability which must, for all practical purposes, be taken as random and unpredictable. The formulation of Eq. (1) simply says that the long-term average duration of recovery of the i th pacer ($\overline{X}_{b,i}$) is subject to cycle-by-cycle modification in an unpredictable manner; and that the Gaussian distribution can be used, together with the parameter α, to evaluate the probability of any particular duration of recovery in a given cycle — for example, about 95% of the recovery phases will be in the range, $\overline{X}_{b,i} \pm 2\alpha$. Note that the *rate* of the recovery process, which plays a central role in models for high-frequency spontaneous neurons and which underlies the several alternative formulations for these models, becomes an implicit rather than an explicit variable in the formulation of Eq. (1).

The pacers of the ensemble are assumed to be not only individually erratic, as quantified by α, but also intrinsically somewhat different from each other in their average frequency: different in the rates of recovery. In order to incorporate such differences into the value of the duration of recovery, with a single additional parameter, the Gaussian distribution is again convenient; let

$$\overline{X}_{b,i} = \overline{X}_b + \beta z_{i,\beta} \tag{2}$$

where, as before, $\overline{X}_{b,i}$ is the average duration of the recovery phase for the i th pacer in the basic state; $\overline{X}_b$ is the overall average of the recovery phase, in the basic state, for the entire population of pacers, the mean of the means; $z_{i,\beta}$ is a normally distributed random variate (mean of zero, variance of one), an intrinsic and invariant property of the i th pacer; and β is a coefficient which converts pacer-to-pacer variability into hours: the standard deviation of the distribution of *average* recovery phases. The average duration of recovery for a given pacer in the basic state thus remains constant; for a pacer selected at random, the average duration will, with about 95% probability, be in the range, $\overline{X}_b \pm 2\beta$.

While both Eq. (1) and Eq. (2) invoke the Gaussian distribution, it should be emphasized that very different issues are involved in the uses to which it is put. The Gaussian variate z_j in Eq. (1) deals with temporal unpredictability, and reflects dynamic, physiological "sloppiness" of the single pacer. The Gaussian variate $z_{i,\beta}$ in Eq. (2) deals, on the other hand, with the diversity of intrinsic properties within the ensemble, with enduring differences among the pacers which can be taken to reflect the equivalent of intercellular differences in anatomy.

The recovery phase is only one of the two portions of a complete cycle of a pacer. In order to define the duration of the other portion, the discharging phase, the simplest course might well be to invoke three additional and similar parameters: one for the mean duration of discharge, one for pacer-to-pacer variability in that duration, and a third for cycle-to-cycle stochastic variability. Instead, however, I have relied primarily upon a two-parameter formulation, which makes the duration of discharge directly proportional to the preceding recovery phase:

$$Y_{i,j} = \delta_i X_{i,j} \tag{3}$$

where $Y_{i,j}$ is the duration of the discharge phase of the i th pacer in the j th cycle, $X_{i,j}$ is the duration of the immediately preceding recovery phase of that pacer; and δ_i is a dimensionless coefficient of proportionality, specific to the i th pacer. Note that $Y_{i,j}$ and $X_{i,j}$ in [Eq. (3)] have no subscript referring to the basic state. This is intentional, and implies that [Eq. (3)] is general, in the sense that it is not restricted to the basic state in which each pacer acts independently; the formulation also applies to the pacers when they are coupled, as described subsequently. Then, in order to incorporate pacer-to-pacer differences in Y, let us again invoke the Gaussian distribution, as follows:

$$\delta_i = \overline{\delta} + \gamma z_{i,\gamma} \tag{4}$$

where $\overline{\delta}$ is the overall average coefficient of proportionality for all pacers; $z_{i,\gamma}$ is a normally distributed random variate (mean of zero, variance of one), an intrinsic and invariant property of the i th pacer [assigned independently of $z_{i,\beta}$ in Eq. (2)]; and γ is a scaling factor to convert the variance into units commensurate with $\overline{\delta}$. About 95% of the values of δ_i will thus lie in the range, $\overline{\delta} \pm 2\gamma$.

An intuitive argument in favor of the proportionality of Eq. (3) can be made by speculation about what is happening during the recovery process: if some substance is being accumulated or synthesized, then an unusually long duration of recovery in a given cycle might lead to greater storage of the substance accumulated, thereby perhaps permitting subsequent discharge to be longer. A further advantage of this formulation is that δ_i also serves, in part, as a pacer-specific multiplier for α. It thereby introduces a pacer-to-pacer variation in the extent to which stochastic processes affect cycle length, meaning that no two pacers will have identical cycle-to-cycle variability in total cycle length.

As will be, however, subsequently demonstrated (Chap. 6), the assumption that the discharge phase of a cycle is proportional to the preceding recovery phase is by no means critical to the kinds of models of interest. Alternative formulations for $Y_{i,j}$, involving other kinds of parameters, will be considered, and those alternatives apparently do not cause the pacemaker to function differently in any significant way.

Operationally, in order to define the full cycle length of a given pacer in the basic state, with both recovery and discharge phases, the first step will be to define the "anatomy" of that pacer, by assigning to it randomly chosen values of the Gaussian variates, $z_{i,\beta}$ and $z_{i,\gamma}$. The full cycle is then also subject to a cycle-specific random variability, determined by z_j. Combining Eqs. (1) to (4), the total duration of a single cycle can be described as follows:

$$\tau_{b,i,j} = X_{b,i,j} + Y_{b,i,j} = (\overline{X}_b + \alpha z_j + \beta z_{i,\beta})(1.0 + \overline{\delta} + \gamma z_{i,\gamma}) \tag{5}$$

where $\tau_{b,i,j}$ is the cycle length, i.e. the interval between successive onsets of the recovery phase, of the i th pacer in the j th cycle, in the basic state. The duration of each successive cycle of that pacer can be also be determined with Eq. (5), simply by drawing at random in each cycle a new value for z_j from the unit normal distribution.

Although it may not be evident at first glance, Eq. (5) is a way of defining the cycle length of an elementary kind of system known as a relaxation oscillator. The essence of this sort of system is that some substance or property accumulates up to a critical level, at which time a reversal process begins and proceeds to completion, and the accumulation process then begins anew. An elementary deterministic scheme of this sort is illustrated in Figure 4.2A, and the addition of random noise to the system is illustrated in Figure 4.2B. The analogies with the models for high-frequency neurons illustrated in Figure 4.1 are self-evident. For present purposes, the instantaneous value of the recovery process of a pacer shown in these illustrations

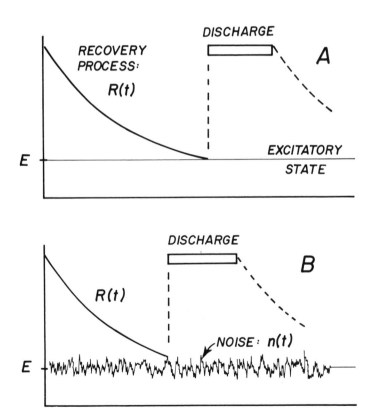

Fig. 4.2 A and B. Schematic illustration of modification of the spontaneous-neuron models, as proposed for a pacer, in elementary form (part A), and with stochastic variation incorporated (part B). At the end of pacer discharge, a recovery process, $R(t)$, begins. When prevailing excitatory state, E (or E plus stochastic variation, $n(t)$) exceeds $R(t)$, pacer discharge is initiated, and thereafter proceeds to completion

is only implicit; interest focusses instead upon the durations of the recovery phase and the discharge phase, as these are specified in Eqs. (1) to (4).

Note that in a system of this sort, there is no cycle-to-cycle "memory" of preceding behavior. The pacer returns to an identical state at the end of each discharging phase; there will therefore be no correlation whatever between its performance in successive cycles. This particular property of the pacers deserves emphasis because, as will be shown later (Chap. 12), the ensemble behavior turns out to be appreciably more complex, no longer resembling that of a relaxation oscillator. The pacemaker system as a whole can show evidence of memory which persists over many circadian cycles.

While the illustration of Figure 4.2B suggests that the pacers envisioned here are very elementary in their dynamics, the quantitative specification of that behavior in Eq. (5) is a lengthy one. Five parameters (α, β, γ, δ, and

Table 4.1. Parameters of the pacemaker model

Symbol	Referent	Dimensions	"Typical" value
$\overline{X}_b$	Mean duration of recovery: Overall, long-term average duration of the recovery process of the general ensemble of pacers	Hours	12 to 20
α	Stochastic coefficient: Standard deviation in the duration of the recovery process	Hours	0.5 to 3
β	Between-pacer variability: Standard deviation of the variability between pacers in the average duration of the recovery process	Hours	0.5 to 3
$\overline{\delta}$	Discharge coefficient: Coefficient of proportionality between duration of recovery process in a given cycle and duration of subsequent discharge phase	Dimensionless	0.2 to 0.6
γ	Discharge variability: Standard deviation of the coefficient of proportionality relating duration of recovery and discharge phases	Fraction of $\overline{\delta}$	$\overline{\delta}/20$ to $\overline{\delta}/4$
N	Number of pacers: Size of the pacer ensemble	Dimensionless	10 to 800
ϵ	Excitation: Maximum extent to which discriminator feedback can shorten the recovery phase of a pacer	Hours	4 to 12
θ	Threshold: The number of pacers which must be discharging simultaneously for their activity to produce feedback, which shortens the recovery phase of pacers	Fraction of N	N/10 to N/2

$\overline{X_b}$: see Table 4.1) have already been invoked; a sixth parameter, the number of pacers in the ensemble (to be designated N), is implied; and still no word has been said about coupling, the kind of interaction which proves to be the really interesting property of the system. The wary reader may, at this point, begin to suspect that the model system under consideration shows embryonic symptoms of developing into an amorphous monster: one which can, like an ameba, engulf and account for any set of data by modification of one parameter or another, but which thereby merely demonstrates its lack of backbone. Can any model with so many parameters lay legitimate claim to being a "simple" model?

The search for simplicity in an explanation or model always entails the risk of oversimplifying the data for which one hopes to account. The problem arises in deciding at what point a given body of experimental data has been adequately explained by a model: which features of observation should be emphasized and which others may be overlooked as irrelevant, or at least unimportant? In the preceding chapter, a central issue of this sort has arisen, about which legitimate differences of opinion can and do exist. Because the precision observed of the circadian wake-sleep cycle is commonly so great, previous descriptions of the rhythmicity have been undertaken almost exclusively by invoking deterministic models, which state that at exactly time "t", event "y" must occur. While the issue of temporal variability in model systems for circadian rhythms has not been completely ignored by others (cf. Aschoff et al. 1971), no prior model has given detailed consideration to a mechanism by which the observed precision might arise. This astonishing property of the rhythms is thereby treated as a peripheral issue, one which is to be taken for granted, rather than one demanding explanation. Perhaps there are, in fact, single cells in the brain of a bird or mammal which are, by themselves, able to generate extremely precise circadian rhythms. If one accepts this premise, the problem of how the temporal precision arises would of course remain: by what mechanism could a single cell be so precise? One could, however, thereby ignore that question in formulating a model for the performance of the whole animal, by proposing that it is an issue of intracellular dynamics which need not be of immediate concern.

In contrast, my approach is to focus immediately upon precision as a central issue: in view of the large measure of variability inherent in most other biological rhythms, how can the circadian wake-sleep rhythm be so precise? Rather than gloss over circadian precision with the assumption of extremely reliable single cells, I choose to confront this issue head on, and to deal with it by the suggestion that it might be possible to assemble a reliable pacemaker out of an ensemble of neurons which are basically very sloppy. A necessary consequence of this point of departure is that the pacemaker model must deal with many components, and incorporate parameters to quantify their dissimilarities and their unreliability. Additional assumptions and parameters will be required to quantify the interactions of the ensemble.

From an elementary understanding of the idea that a simpler model is a better model, such an approach may seem to be misguided; a deterministic model is very apt to be simpler to describe and understand. As an alternative approach to simplicity, however, one may decide, as I have, that it would probably be appreciably simpler, in the workshop of natural selection, to assemble a reliable pacemaker out of components which are sloppy, i.e., erratic and somewhat dissimilar, than to demand that the basic construction elements all be identical and fully predictable in behavior.

If one ignores this issue, the behavior of a deterministic pacemaker, consisting of only one pacer, could be quantified with a single parameter, the equivalent of $\overline{X}_b$. Even if one grants the need for two distinct phases per cycle in a model for the wake-sleep cycle, a two-parameter specification would be possible, invoking parameters which are the equivalent of X_b and Y_b (or, alternatively, X_b and $\overline{\delta}$). To concede that stochastic variation deserves to be considered an "important" property of a pacer, however, is to acquiesce to the introduction of a third parameter, namely an equivalent of α.

As a next step, it seems unrealistic to suppose that only a single neuron is involved in generation of the wake-sleep cycle. Surely this important a responsibility would not be entrusted to only one cell, and if not, we are led to consideration of a multi-pacer system. For those who seek only to reduce the number of parameters in a model, the accession to this suggestion must seem to open Pandora's box. First, of course, the number of pacers must then be specified: another parameter required. Then the issue of possible inter-pacer variability in average performance arises. If one has granted the desirability of stochastic variation in the cycle-to-cycle behavior of a single pacer, how can one resist the realistic expectation that no two pacers are apt to be completely identical in their long-run average behavior? The developmental processes leading to a multi-pacer system might well produce an ensemble of units which are more or less similar in properties — but absolutely identical? Unlikely!

To define differences between pacers in their properties requires at least one more parameter, but into which phase of the cycle (recovery or discharge) should that parameter be incorporated? Further, is it plausible that the other phase of the cycle would be completely without variability between pacers? And while we're at it, is it plausible that the influence of stochastic variation would be identical for each pacer? The formulation of Eq. (5) represents a compromise solution to these questions: by invoking two additional parameters (β and γ) the model incorporates inter-pacer variation into each phase of the cycle, as well as into the stochastic element.

The fact that the specification of the model already involves six parameters may well be regarded as a disconcerting state of affairs, but much of that apparent complexity is a necessary outgrowth of the fact that we have undertaken to deal with an ensemble of imperfect oscillators. In such a system, it is possible to dispense with $\overline{\delta}$, in a special case considered subsequently which proves to be largely uninteresting (Chap. 8); and γ can be readily omitted from the models, at a serious cost in realism; but parameters which are the conceptual and functional equivalents of $\overline{X}_b$, α, β and N seem to be logically inescapable.

4.1 Structure and Dynamics of the Coupling

With this extensive prelude, defining the behavior of the pacers in the independent or basic state, we are now in a position to deal with the central issue of this chapter: the mode of interaction among the pacers, by means of which precision of the ensemble behavior might be enhanced. As a foundation for quantitative specification, the schematic diagram of Figure 4.3 proposes a kind of structural arrangement of the pacemaker which

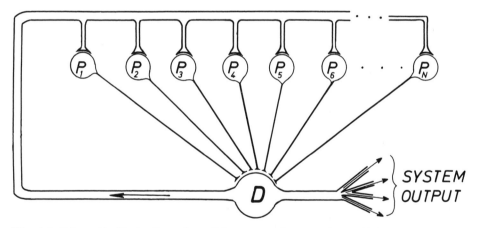

Fig. 4.3. Schematic illustration of possible structural connections which might mediate the kinds of interactions between pacers (P_1, P_2 ... P_N) and the discriminator (D) envisioned in the pacemaker models

might underlie the kinds of interactions proposed. In this scheme, each pacer of the ensemble transmits its output directly to a "master" unit, designated a "discriminator", and receives its only input through another channel which arises from the discriminator. The discriminator is sketched as consisting of a single cell, but its function could well be subsumed by a group of cells, perhaps even those of an entire nucleus in the brain which, through interconnection, produces a unitary output. The essential feature of this diagram is that the individual pacers of the ensemble are not directly interconnected each to the other. Interaction is possible, via the indicated synaptic connections, but the effects of one pacer on another are subject to modification, being contingent upon the state of the discriminator.

The discriminator is envisioned as a completely passive element in the system, an integrator with no long-term memory, no inherent rhythmicity. Its function is comparable to that of a spring-loaded on-off switch; if (and only if) its present input exceeds a given value, it generates a fixed output. Input from a pacer to the discriminator is assumed to arise only when the pacer is in its discharging phase. Each discharging pacer contributes a value of one to a sum, and the instantaneous value of this sum, from the entire ensemble of pacers, is the stimulus — for the time being, the only stimulus — affecting the discriminator. Should the sum be less than a critical value called threshold (designated by the parameter θ), the discriminator remains in its inactive state, during which it produces no output. In the absence of discriminator output, the individual pacers have no effect on each other; they are in their basic state, and behave completely independently, as specified in Eqs. (1) to (5). When the input sum exceeds θ, the discriminator "switches on" and produces feedback — here envisioned as a continuous excitatory train of nerve spikes — which has the potential of affecting any pacer not already discharging. A central assumption about the mode of interaction envisioned here is that as an excitatory stimulus, discriminator feedback can only accelerate a pacer; it can potentially terminate the recovery phase, inducing premature onset of the discharge phase.

This proposed mode of action is based on direct homology with models for high-frequency spontaneously active neurons (Fig. 4.1). In these models, an increase in tonic excitatory input is envisioned as raising the "excitatory level", and thereby shortening the interval between successive spikes (Fig. 4.4). For the circadian pacers, the stimulatory effect of discriminator output is also envisioned as raising the excitatory level, at which the recovery process is terminated and the discharge phase is initiated (Fig. 4.5). As in the comparison of Figures 4.1 and 4.2, the only significant difference proposed between the two kinds of system in Figures 4.4 and 4.5 is that a single spike is initiated in the high-frequency neuron, and a several-hour burst of sustained discharge is triggered in the circadian pacer.

The extent to which discriminator activity modifies the behavior of the pacers of the ensemble constitutes the last element of the pacemaker models which must be quantified. The magnitude of this effect is determined by the difference between excitatory levels, i.e., $E_1 - E_2$ in Figure 4.5B, but for calculational purposes it is more convenient to express the effect

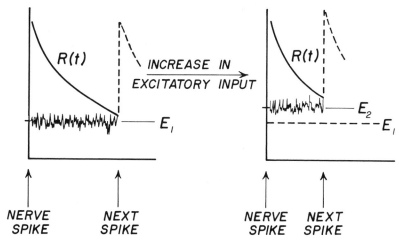

Fig. 4.4. Schematic illustration of the consequences of increased levels of tonic excitatory input in models of the sort shown in Figure 4.1. An increase in excitatory state, E, shortens the average interval between successive nerve spikes

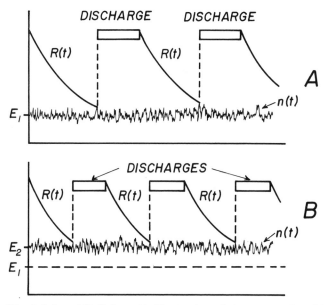

Fig. 4.5 A and B. Schematic illustration of the manner in which discriminator feedback is envisioned as affecting a pacer. The resulting increase in excitatory state, E, shortens the duration of the recovery phase, in a manner comparable with that in which tonic stimuli affect the spontaneous-neuron models, as illustrated in Figure 4.4. **A** Behavior of a pacer in its basic state; **B** The accelerated state

in terms of the amount by which the recovery process of the pacer is shortened by discriminator activity. When the discriminator is continuously active, and continuously exerting this excitatory stimulus, each pacer is

accelerated, and the amount of this acceleration constitutes the final required parameter, ϵ, which functions as follows:

$$\overline{X}_a = \overline{X}_b - \epsilon \qquad (6)$$

where, as before, $\overline{X}_b$ is the ensemble mean duration of recovery in the basic
(i.e., independent) state; $\overline{X}_a$ is the ensemble mean duration of recovery in
the accelerated state (i.e., when receiving stimulation from the discriminator); and ϵ is the difference in mean duration in hours.

4.2 Intermittent Feedback

The general kind of model to be considered here is, in principle, now
completely specified, and involves eight parameters: the six used to define
the ensemble of pacers in their uncoupled state, plus discriminator threshold, θ, and the steady-state effect of continuous discriminator feedback,
ϵ (Table 4.1). There is, however, a calculational matter to be dealt with,
which has important effects on the performance of the system. The application of Eq. (6) to the pacer, during prolonged, continuous activity of the
discriminator, is straightforward, but intermittent feedback from the discriminator presents potential complications. For example, how does one
deal with the behavior of a pacer which begins its recovery phase while the
discriminator is inactive, and which is then subject to a brief interval of discriminator feedback, which then terminates before the pacer's recovery
phase has ended? The calculational procedure must be clarified.

The manner in which such contingencies are handles here is also a direct
outgrowth of concepts which originate from models for high-frequency
spontaneous neurons. In those models, the recovery of sensitivity is regarded
as an intrinsic process which proceeds at the same rate regardless of stimulation levels. Similarly with the circadian pacers, the recovery process (as
well as the discharge) are presumed to be unaffected in their dynamics by
feedback from the discriminator; only that excitatory level at which recovery terminates is altered by discriminator activity. Hence, as shown in
Figure 4.6, a brief interval of discriminator input to a pacer *may* shorten
the cycle of a pacer; and whether it does so or not depends upon *when* the
input occurs. Another way of illustrating this mode of action is contained
in Figure 4.7, dealing with very brief pulses of discriminator activity.

In the computer program for simulations of system performance (App.
4.B), the changes in excitatory level of the pacers, illustrated in Figure 4.6,
do not appear explicitly. They are, however, implicit in the manner in which
onset and offset of discriminator activity during the recovery phase of a
pacer are handled. At the moment of discriminator onset (or offset), the
entire probability-density function associated with Eq. (1) [which must now

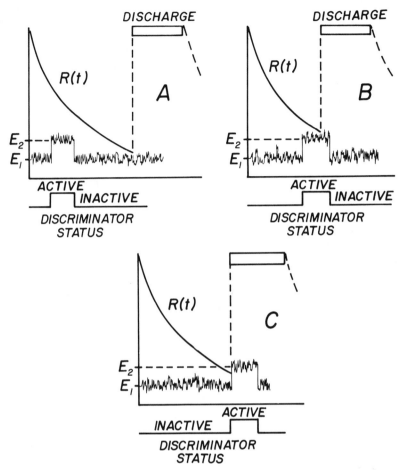

Fig. 4.6 A−C. Schematic illustration of the consequences of short-duration pulses of discriminator feedback upon a pacer, at various times during the recovery process. **A** a stimulus early during recovery is without effect on the pacer; the result is an unaltered or basic cycle; **B** a stimulus later during recovery leads eventually to shortening of the recovery interval; the resulting recovery interval will be equivalent to that expected had the discriminator been continuously active, that is, a fully accelerated cycle; **C** a stimulus still later during the recovery leads to an almost immediate onset of discharge; the duration of recovery in that cycle would be intermediate between that seen in parts **A** and **B** of this figure: a partially accelerated cycle

be specified as a conditional probability: given that a discharge has not yet begun, what is the likelihood that it will begin in the next unit of time? See App. 4.A] is shifted to an earlier time (or a later time, for discriminator offset) by the full amount of ϵ hours.

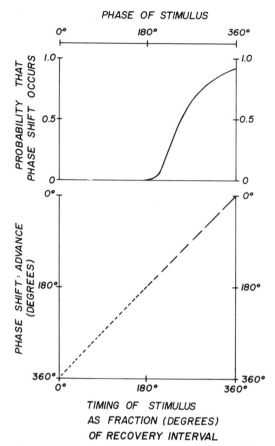

Fig. 4.7. Effect of a single very brief pulse of discriminator activity on recovery phase of a pacer, as a function of timing of the stimulus within the cycle. Because the pacer is a simple relaxation oscillator, the magnitude of the predicted phase shift is always exactly equal to the extent by which the stimulus precedes expected spontaneous discharge. The magnitude of feedback, specified by the parameter ϵ, alters only the *probability* that such a phase shift will occur; it affects only the position of the probability curve along the abscissa. The shape and curvature of the probability curve depend upon the parameter α. In some senses, the interval where probability is essentially zero corresponds to an absolute refractory time, and the initial nonzero portion of the curve – up to, say, $p = 0.5$ – then corresponds to a relative refractory time. For a stimulus of double this duration, the probability can be calculated by the usual rules of combining probabilities ($p_1 + p_2 - p_1 p_2$), and the amount of phase shift is no longer a line of unit width, but a band, extending upward from the 45° line, two duration-of-stimulus units wide

4.3 Discussion of the Coupling

In the specification of how mutual interaction among pacers occurs, the formulation chosen is obviously conservative, in the sense that only two additional parameters were invoked, one to indicate the necessary condition for application of feedback (θ), and the other to indicate the magnitude of feedback (ϵ). A far more important issue than the number of required parameters, however, is the rationale which underlies this postulated mode of interaction. The nature of the effect of feedback assumed here is a critical characteristic of the kind of models under consideration; it serves to distinguish these models from all others; and it therefore demands very careful scrutiny.

There are two key elements in this treatment; the first of these is that the effect of the pacemaker ensemble upon any single pacer can be only to shorten its intrinsic basic cycle, never to lengthen it. This is equivalent, in terms of the nervous system, to the assumption that the discriminator can provide an excitatory stimulus but never an inhibitory one. The process of mutual entrainment would be a very different one, if one assumed that a given stimulus can either speed up or retard the cycle of a pacer, depending upon phase within the cycle at which the stimulus is administered: a cyclically varying sensitivity of the pacer. Many nonbiological oscillators operate in this latter way, and that sort of formulation underlies several previous treatments of mutual entrainment of populations of oscillators (Winfree 1967; Pavlidis 1969, 1971). Furthermore, as we shall see subsequently, the circadian wake-sleep rhythm of a bird or a rodent can be either accelerated or retarded by a light stimulus, depending upon when within the circadian cycle the stimulus is administered. In view of the neurophysiological literature, however, I believe that such variation in sensitivity of a biological rhythm to phase-shifting stimuli is an unusual property of the pacemaker rhythm which must be *accounted for,* rather than being taken for granted as an intrinsic property of the constituent neuronal oscillators.

It is well known, of course, that some kinds of stimuli can act in an excitatory mode, increasing firing frequency of a receptor; and that others can act in an inhibitory mode, decreasing frequency. Cases are also known in which the same stimulus can act as either excitatory or inhibitory, depending upon the nature of the receptor and the synapse through which the input is transmitted; but it is a very different matter to suggest that a given stimulus can act on a single neuron as either excitatory or inhibitory, shortening or lengthening cycle duration, depending only upon *when* within the receptor cycle the stimulus is administed. There are isolated instances in which data suggesting such unusual behavior have been reported for particular neuronal preparations (Ayers and Selverston 1979; Fohlmeister et al. 1974; see, however, Hartline 1976), but the far more ordinary situation is that a stimulus acts only unidirectionally upon a receptor rhythm, granted, of course, that the *amount* of acceleration by the stimulus (or delay, with inhibitory stimuli) often depends upon when it is given.

Because this sort of unidirectional responsiveness is the norm in the behavior of high-frequency, single-neuron rhythms, and because the models for circadian pacemakers proposed here purport to deal with neuronal systems, I have relied almost exclusively on unidirectional (cycle-shortening) effects of the discriminator on the pacers envisioned here. It would be entirely consistent with neurophysiological data, of course, to postulate that the discriminator has an excitatory effect upon some pacers and an inhibitory effect upon others, but as we shall see, this kind of additional complexity in the models proves unnecessary.

The second key element in the treatment of discriminator action is embodied in the schematic diagrams of Figure 4.6: the assumption that the discriminator activity does not affect the *rate* of the recovery process, but only alters that critical level at which recovery terminates and discharge begins. This interpretation is not simply a matter of arbitrary assumption, but was incorporated in the models because it follows naturally from models described above (cf. Fig. 4.1), which have been proposed to account for the responses of spontaneously active high-frequency neurons. Those models in their elementary form cannot adequately describe the large body of data on single-neuron behavior without such an interpretation (Enright 1967). There is, of course, no compelling reason to insist that neurons which have cycles of many hours in duration will necessarily behave similarly, but at the least, this mode of stimulus action appears to me to be a sensible point of departure for models which purport to deal with interactions among neurons.

It is worth emphasis that in the treatment here the discriminator remains a completely passive element, responding only to prevailing input; and that there is no temporal variation in the way in which its feedback affects the pacers. Whenever the discriminator becomes active, it raises the excitatory level of each pacer by a fixed amount (Fig. 4.6). This is handled, in the calculations, by a discrete shift in the conditional-probability distribution, which relates to completion of a pacer's recovery phase, toward a potentially shorter cycle length. The magnitude of that shift in the probability distribution is completely invariant with time in the cycle of a pacer, and is described by the parameter ϵ.

Nevertheless, as illustrated in Figures 4.6 and 4.7, the actual consequences of this mode of action can be somewhat complex. When a pacer is still discharging, or has just begun its recovery phase, a brief pulse of feedback from the discriminator will have no effect whatever on realized cycle length: as though the pacer were in a state equivalent to the absolute refractory time of a neuron (Fig. 4.6A). If the pacer has nearly achieved the time of its spontaneous onset of discharge, then sudden onset of discriminator feedback will almost certainly trigger an immediate premature onset of pacer discharge (Fig. 4.6C): a reliable, reflex-like behavior typical of many kinds of neurons in response to strong phasic stimuli. If the pacer is midway into its recovery phase, onset of discriminator action may or may

not trigger premature onset of discharge (analogous to the relative refractory time in an ordinary neuron); a large stimulus (i.e., large value of
ϵ) will trigger discharge but a small one will not. If it does not (that is, if
the pacer continues only to recover, and the discriminator then switches
back to inactivity), then the pacer cycle remains completely unaffected by
a pulse from the discriminator (Fig. 4.6B). This kind of action is easy
enough to describe graphically as in Figure 4.6, and follows naturally from
elementary models for high-frequency neurons; it is straightforward in concept, but is a sort of interaction which has not, I believe, been previously
considered in the mathematical formulation of coupling among oscillators.

Perhaps an analogy will assist in visualizing the process involved. If an
alarm clock (a relatively quiet one) begins buzzing shortly after a person
has fallen into a sound sleep, and stops after a few minutes, awakening is
very unlikely (Fig. 4.6A); if it begins to buzz just shortly before spontaneous awakening would occur, in the early morning hours, during very light
sleep, then it is apt to jar the sleeper into immediate wakefulness (Fig. 4.6C);
if it begins somewhat earlier, the person may or may not awaken, depending on how loud the noise is, but if not, and the alarm then stops after a
few minutes, the sleeper would continue thereafter to slumber on through
to his normal, spontaneous awakening (Fig. 4.6B). The excitatory effect
of the discriminator on a pacer resembles such effects of an alarm clock.
This mode of action gives the models here a variety of interesting and hitherto unexplored consequences and properties.

Although the basic mode of coupling among pacers proposed here derives
from extrapolations of the known behavior of high-frequency spontaneously firing neurons, there are a number of ways in which the specific formulation proposed for these effects may appear to be unrealistic: in some ways
too simple, in others perhaps too complex. Some of the legitimate questions
which might be raised in the name of "realism" include the following:

1. An appreciable number of pacers, specified by θ, must be simultaneously discharging before the discriminator becomes active. Would not the system be simplified, if the discriminator produced feedback as soon as any
single pacer begins to discharge?

2. The discriminator acts only as an on-off, all-or-none switching element.
Would not a quantitative, continuous relationship — perhaps even direct
proportionality — between input and output be more realistic? (This question is in fact another way of asking how important it is that the discriminator be a separate, discrete element of the pacemaker.)

3. The discharges of the separate pacers in the ensemble are summed as
a simple, arithmetic operation, and this sum is compared with threshold,
θ. Is such simple addition plausible in a neuronal system?

4. The discharge of each pacer is treated as entirely equivalent, in calcu-
lating that sum, and independent of time during the discharge phase. Should
one not expect inter-pacer variability in these factors?

5. The effect of discriminator feedback is taken as identical (ϵ hours) for each pacer in the ensemble. Should not inter-pacer variability be incorporated into this parameter?

6. Much has been said about a precise system made of sloppy components, but no stochastic variability has been assigned to the threshold of the discriminator. Is this realistic?

Each of these issues will be subsequently examined in some detail, but before such legitimate concerns are attended to, it is essential to develop an understanding of how the elementary coupled system, as proposed, behaves under steady-state conditions.

Appendix 4.A. Calculations Based on Conditional Probability

The calculational procedure for dealing with stochastic behavior of a pacer during coordinated system performance is more complex than Eq. (1) implies. That equation would be entirely adequate, during long-term absence of discriminator feedback: at the start of the recovery phase, one would simply select at random a new, cycle-specific value of the Gaussian variate, z_j; and during long-term continuous feedback, only the modification of Eq. (6) would be required. If the discriminator, however, switches on or off during a pacer's recovery phase, an alternative culculational procedure involving conditional probability becomes necessary. Consider, for example, the case in which ϵ has a value of 8 h, and in which the value of z_j for a particular cycle dictates that the recovery phase of a given pacer should be 18 h, if the discriminator is inactive; with ϵ being 8 h, the duration should be 10 h if the discriminator is active. Assume that the discriminator is inactive at the start of the recovery phase, and then begins to produce feedback 16 h later. Neither 10 nor 18 h is the appropriate duration of recovery for that pacer in that cycle, nor would it be correct to decide that the pacer must begin discharge as soon as feedback had begun, although the probability should be high that recovery would terminate very soon after onset of feedback. In order to handle this situation properly, the required calculation must be based upon the likelihood that recovery would terminate in the next sequential series of time units — given that recovery has not yet ended.

Under steady-state conditions (discriminator continuously inactive or continuously active), the probability that discharge of a pacer will begin in any given time unit, Δt, which differs from the mean by a value of αz_j, is approximately

$$\int_{z_j}^{z_j + \Delta t} y \, dt \tag{A1}$$

where y is: $[\exp(-z_j^2/2)]/\sqrt{2\pi}$. This situation can be handled without attention to the probability in each time unit simply by selecting at random a value of the ordinary Gaussian variate, as proposed in Eq. (1). If the discriminator, however, can suddenly begin to produce feedback, midway into a recovery interval, in the manner described above, we must return to the fundamental equation, and modify it to allow for the fact that recovery has not, in fact, terminated in any prior time interval. The appropriate correction factor leads to the following expression for the probability that discharge begins in any particular time unit:

$$\int_{z_j}^{z_j + \Delta t} y \, dt \left/ \left(1.0 - \int_{-\infty}^{z_j} y \, dt\right)\right. \qquad (A2)$$

where all symbols are those of Eq. (A1), and where this equation must be applied sequentially to all time units after onset of recovery. Thus the simple Gaussian probability is converted into a conditional probability, which is uniquely determined by the mean duration of recovery, the time since last onset of recovery and the parameter α.

For pacers acting independently in the basic state, this rather complex calculational procedure serves only as an indirect and lengthy way of reconstructing the equivalent of a randomly chosen Gaussian variate from a distribution with standard deviation α. During intermittent application of feedback, the procedure serves as a way of shifting the entire probability-density function of interval lengths forward or backward in time, with appropriate correction for events that have not taken place.

Appendix 4.B. A FORTRAN Program for Computer Simulation

In order to deal with cycle-to-cycle stochastic variability, we will truncate the Gaussian distribution at $\pm 6\,\sigma$, and formulate the distribution in terms of conditional probability. After choice of an arbitrary time unit (ATU) as the basis for calculations (1/4 h was used in all calculations to be described here), let A = one hour/ATU, and select ALPHA [α of Eq. (1) in text] in hours. We will need 12*A*ALPHA (= M) values for a conditional probability function, P(I), which specifies the probability that a pacer begins to discharge in time unit I, granted that it has not begun to discharge up to that time. Let STEP = ATU/ALPHA. As I assumes integer values between 1 and M,

$$P(I) = \frac{\int_{-6 + I*STEP}^{-6 + (I + 1)*STEP} \frac{1}{\sqrt{2\pi}} \exp(-x^2/2)dx}{1.0 - \int_{-\infty}^{-6 + I*STEP} \frac{1}{\sqrt{2\pi}} \exp(-x^2/2)dx} \tag{A3}$$

Now select values, in hours, for BETA, XBAR and EPSIL [β, $\overline{X}_b$ and ϵ in Eqs. (2) and (6) in text], values for DELAV and GAMMA [δ and γ in Eq. (4) in text], and values for NCELL (the number of pacers), THR (threshold, in units of number of pacers) and ITIME (the length of the desired calculation, in number of ATUs). Let GAUS be a unit random normal variate (mean = 0, variance = 1.0), generated anew each time called; and let RANDM be a random number from a rectangular distribution, with a range from zero to one, generated anew, each time called.

The following listing indicates the required sizes of the arrays to be stored: stored: XMIN(NCELL), DELTA(NCELL), ISET(NCELL), P(M), JA(ITIME), SUM(ITIME). The following program is sufficient to simulate performance of the system, after input data which specify BETA, XBAR, EPSIL, DELAV, GAMMA, NCELL, A, THR and ITIME, as well as P(I) for the array from 1 to M.

```
 1 DØ 4 ICELL = 1,NCELL
 2 XMIN(ICELL) = A*(XBAR + BETA*GAUS − 6.*ALPHA)
 3 DELTA(ICELL) = DELAV + GAMMA*GAUS
 4 ISET(ICELL) = 0
 5 JA(1) = 0
 6 DØ 23 J = 2, ITIME
 7 J1 = J − 1
 8 SUM(J) = 0.
 9 DØ 18 N = 1, NCELL
10 IF[J − ISET(N)] 17,17,11
11 I = J − ISET(N) − FIX[XMIN(N)]
12 IF[JA(J1)] 14,14,13
13 I = I + FIX(EPSIL*A)
14 IF(I)18,18,15
15 IF[P(I) − RANDM] 18,16,16
16 ISET(N) = FLT(J) + (FLT[J − ISET(N)])*DELTA(N)
17 SUM(J) = SUM(J) + 1.
18 CØNTINUE
19 IF[SUM(J) − THR]20,20,22
20 JA(J) = 0
21 GØ TØ 23
```

22 JA(J) = 1
23 CØNTINUE

The array of values, SUM(J), J = 1, ITIME, constitutes the number of pacers which are in their discharge phase at any time; the array of values, JA(J), J = 1, ITIME, is an index of whether the summed discharge exceeds discriminator threshold, leading to discriminator activity (value of 1) or not (value of 0). If the influences of light are to be considered in the simulation (Chaps. 7–11), they should be incorporated as elaborations on statement 19.

Chapter 5
A "Type Model" and its Behavior: Partial and Loose-Knit Mutual Entrainment

In the equations of the preceding chapter, several quantitative character-istics of the pacemaker system have been left unspecified, being represented only by symbols which were referred to as "parameters". The equations therefore define an entire class of models. In order to examine the behavior of any particular model of this class, numerical values must be assigned to the parameters. In this chapter, detailed attention will be focused upon the performance of what I refer to as the "type model", comparable in prin-ciple with the "type specimen" of a taxonomist: a particular, representative instance of the class of models, characterized by a specific set of parame-ter values. In subsequent chapters, variations in performance will be exam-ined, which arise when the parameters are assigned values which depart from those of the type model, so as to convey an impression of the diver-sity of models which are members of the same class.

The parameter $\overline{X_b}$ (Mean Duration of Discharge) is primarily a scaling factor, to which a value of 17 h was assigned for the type model; and $\overline{\delta}$, the Discharge Coefficient, was arbitrarily set at 0.5. Together, these values give the output of the pacemaker a period in the circadian range. β, the Between-pacer Variability in the average duration of recovery, was assigned a value of 3 h. With this value, about 95% of the pacers, when in the basic state, will have average recovery intervals which are within the range, 17 h $\pm$ 6 h ($-2 < z_{i,\beta} < +2$). α, the Stochastic Coefficient which quantifies cy-cle-to-cycle variability, was assigned a value of 2.5 h. (This value is in part associated with a calculational procedure based on an ATU of 15 min; a standard deviation of 10 ATUs, or 2.5 h, permits use of standard tables of the Gaussian distribution, in computer simulation.) With this value for α, about 95% of the recovery intervals for any given pacer will be within about $\pm$ 5 h ($-2 < z_j < +2$) of the average value for that component. A value of 1/8 was assigned to γ (Discharge Variability); with $\overline{\delta}$ of 0.5, this value of γ means that about 95% of the pacers will have discharging phases which are between one quarter and three quarters the duration of the preceding recovery interval. ϵ (Excitation), which specifies the period-shortening ef-fect of discriminator activity, was given a value of 8 h.

An array of examples of the kinds of single cycles of a pacer which might result from these choices of parameter values, for a pacer in the basic state (i.e., the independent mode, without discriminator input), is shown in Figure 5.1.

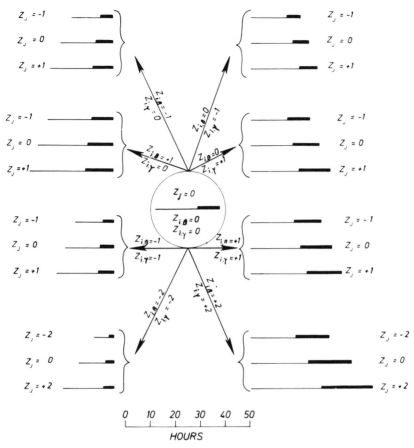

Fig. 5.1. Duration of cycle length for a pacer from the type model, in the basic state, as influenced by various values of z_j, $z_{i,\beta}$ and $z_{i,\gamma}$, the random Gaussian variates of Eqs. (1), (2) and (4). The cycle enclosed in the circle is the "ideal": a typical performance expected as the average over many cycles, from a pacer with properties of the population average (i.e., with all three Gaussian variates set to zero). The examples around the periphery illustrate cycle lengths which would result from different combinations of nonzero values for the Gaussian variates. Differences in z_j, other parameters held constant, reflect the variability in successive cycles of a single pacer; differences in $z_{i,\beta}$ and $z_{i,\gamma}$, when z_j is held constant, reflect differences between pacers in the duration of the recovery and discharging phases, respectively. A value of z_j, $z_{i,\beta}$ or $z_{i,\gamma}$ which is as large as ± 1 should be expected about 1 time in 3; values as large as ± 2 should be expected about 1 time in 20

These examples demonstrate that a very large measure of variability in performance arises from the choices of α, β and γ. Cycle lengths of less than 12 h and of more than 48 h are sometimes to be expected; the idea that the pacers can independently show rhythmicity which is "more or less" circadian in period has therefore been taken very loosely. Inclusion of the "accelerated state" (in which discriminator activity can shorten the recovery phase by up to 8 additional hours) increases even further the range

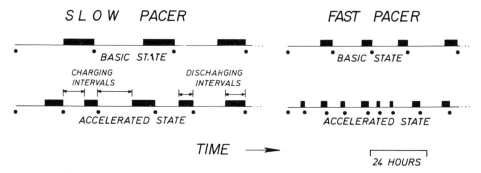

Fig. 5.2. Further examples of the behavior of single pacers in the type model, when feedback from the discriminator is also considered. Between-pacer variability is illustrated by the differences between "slow" and "fast" pacers, which correspond to the case shown in Figure 5.1, when $z_{i,\beta}$ and $z_{i,\gamma}$ both are equal to $+1$ ("slow") or both equal to -1 ("fast"). Variation within the sequences of cycles represents the effect of stochastic variation, z_j. The influence of excitatory input from the discriminator is illustrated by differences between the basic and the accelerated states: since ϵ (Excitation) is equal to 8 h, the duration of recovery of a pacer in the accelerated state will average 8 h shorter than in the basic state. *Dots* beneath ends of discharge phases represent ends of cycles, with reset to zero

of possible cycle lengths for a pacer, as is indicated in Figure 5.2. Since these are all instances of single-pacer cycles which would arise in the absence of coordinated system coupling by the discriminator, they tell us nothing, of course, about behavior or precision of the pacemaker as a whole.

The primary initial justification for making such extreme assumptions about variability in the nature and behavior of the circadian pacers is that one thereby tests robustness of the class of models. If an ensemble of pacers which are as different from each other, and as erratic in their performance, as the behavior shown in Figure 5.1 and 5.2, can still interact to produce a relatively precise system output, then less extreme assumptions about intrinsic variability ought to lead to even better performance.

No empirical measurements are available, of course, which justify the assumption of such large temporal variability in circadian components of the central nervous system of higher vertebrates. In formulating the type model, I was simply making what I believed to be deliberately pessimistic assumptions. Data have recently been published, however, which offer some concrete guidelines for speculation about intracellular and intercellular circadian variability, based on measurements in the unicellular alga, *Acetabularia mediterranea* (Karakashian and Schweiger 1976). Intra-cellular circadian variability, expressed as cycle-to-cycle standard deviation of period over 15 cycles in each of ten isolated cells (ibid. Table 2), had an average value of 2.55 h, with a range from 1.67 to 3.24 h. A portion of this variation was presumably due to measurement error, i.e., imprecision in determining corresponding phase in successive cycles, rather than to intrinsic properties of the cells; confirming this interpretation, eight of the ten measurement series showed negative first-order serial correlation in cycle length. The mean value of the serial correlation coefficient was about -0.16; on the basis of several assumptions (see Pittendrigh and Daan 1976a for details), this coefficient can be used to estimate the mean intrinsic cycle-to-cycle variation in period. That procedure leads to a standard deviation of about 2.1 h: a rather close approximation to the value of the Stochastic Coefficient used in the type model (2.5 h). From the same set of data, two estimates of inter-cellular variation in average period can be calculated, one for each of the two temperatures at which the experiments were conducted. These standard deviations are 1.91 and 1.93 h: estimates which are somewhat less than the value chosen for Between-pacer Variability in the type model ($\beta = 3$ h). To the extent that the *Acetabularia* data can be considered relevant to the problem at hand, these observations suggest that the large variability assigned to pacers in the type model may be a less extreme assumption than I had realized.

In examining the behavior of the type model, we will be considering a range of values for N, the number of pacers present; in one of the examples of this chapter, N was set at 800, and in the other two cases, at 100. For the type model, the threshold of the discriminator, θ, was set at a value of 0.3N, that is, 240 pacers in the first example, 30 in the other examples. To reiterate the function of θ, if less than 0.3N pacers are in their discharge phase, the discriminator is inactive; when more than 0.3N are discharging simultaneously, feedback shortens the average expected recovery interval by ϵ (i.e., 8 h: $\overline{X}_b \rightarrow \overline{X}_a$).

With these values assigned to give the dimensions and properties of the type model, we now have the complete description of a hypothetical pacemaker. For starting conditions, let us assume that all pacers in the ensemble are simultaneously set into motion at the start of their recovery phases. In order to examine in detail how the system behaves, we must turn to computer simulation, but before examining such simulations, certain qualitative expectations can be formulated. Because of the cycle-to-cycle and pacer-to-pacer variability in the recovery interval, $\overline{X}_{b,i,j}$, there will be a very gradual buildup in the number of pacers which are discharging. Once the threshold of the discriminator has been exceeded, however, output from the discriminator will raise the excitatory state of all pacers which are still recovering, switching them from their independent mode and increasing the probability per unit time that they will also begin to discharge. The result of this switchover should be a rapid increase in the number of discharging pacers immediately after threshold, θ, is achieved. The pacemaker ensemble will suddenly experience intense positive feedback, which should result in an almost explosive recruitment of previously quiescent pacers. Data from a computer simulation confirming these expectations are shown in Figure

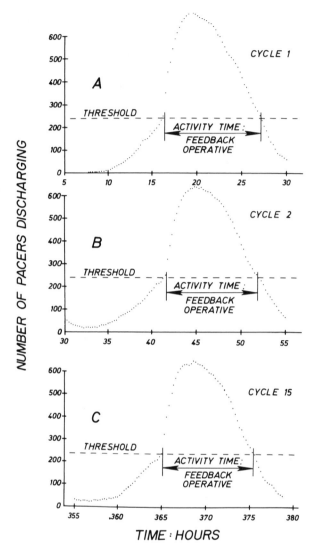

Fig. 5.3 A–C. Excerpts from a simulation with the type model, incorporating 800 pacers in the ensemble. Number of pacers discharging is plotted against time. **A** Cycle 1, beginning with all pacers at the start of their recovery phases; **B** and **C** Cycles 2 and 15, respectively, of a continuation of the same calculation

5.3A; immediately after threshold has been reached (hour 16.25), the pacer-discharge curve climbs very steeply.

If this entire process were to be repeated, starting again with the same set of pacers, simultaneously set into motion at the start of their recovery phases, the attainment of threshold would be qualitatively similar, but different in detail due to random variables (values of z_j). The subset of pacers which would be responsible for the crossing of threshold would again tend to be those with intrinsically short recovery intervals ($z_{i,\beta}$ negative), but it is very unlikely that in a second test they would be the identical units from the ensemble which led to the onset of discriminator activity in the first test. With a finite set of pacers, the time of threshold crossing will also vary from one trial to another. Nevertheless, as simulations have demonstrated, the time of threshold crossing is appreciably more predictable than the performance of any single pacer; and this should not be surprising, since attainment of threshold depends upon averaged performance of the entire ensemble, in which occasional extreme events are "masked".

A calculation of the sort which led to Figure 5.3A would clearly not be representative of the long-term behavior of the system, since it started with the artificial condition that all pacers had been reset to zero, i.e., simultaneously placed at the start of their recovery phases. The behavior of the system during subsequent cycles is more complex. In order to understand this subsequent behavior, note the rapid increase in number of discharging pacers in Figure 5.3A, immediately after threshold crossing. This has created a subset of pacers which are nearly in phase with each other; some 50% of the pacers were more or less synchronized in this manner, within the first 2 h after onset of discriminator activity. This subset represents a group which might be expected, once they have completed their discharge phases, and begun recovery again, to induce another threshold crossing at a later time by a process similar to that in Figure 5.3A.

Further simulations confirm this expectation: Parts B and C of Figure 5.3 show excerpts from a continuation of the calculation which led to Figure 5.3A, during cycles 2 and 15 of the resulting, self-sustained oscillation of the pacemaker. There is no evidence of damping (i.e., of the oscillation dying out in amplitude as time passes), and cycle 15 appears to differ in no significant respect from cycle 2. The primary difference between cycle 1, starting with all pacers reset to zero, and the subsequent cycles is that the first cycle lacks the modest level of "background" discharges which subsequently developed (compare hours 5–10 with hours 30–35 and 355–360).

Because of the assumptions in this class of models, none of the pacers which contributed to the threshold crossing in the first cycle received feedback, or had its timing affected in any way, by the onset of discriminator activity in that particular cycle. Nevertheless, because of the cycle-to-cycle stochastic variability in their performance, a significant fraction of these components would, in the next cycle, fail to begin to discharge before threshold crossing, and therefore would, in cycle 2, be among those which are accelerated by the discriminator. Thus, there will develop an entrained set of pacers which only intermittently receive feedback, but which, *on the average,* will have the same period as the driven oscillations of the discriminator. In any given cycle, only the subset of this group which did *not* contribute to attaining threshold can be affected by feedback. The members of the entrained set must take turns: the subset affected by feedback in a subsequent cycle will have other members. The discriminator maintains coherence of this mutually entrained set of pacers by rephasing them in a sufficient fraction of their cycles to prevent their "escape" from entrainment.

With the large differences present in the average frequency of the pacers of the ensemble, one might suspect that only a fraction of the whole population could be entrained. Some of the pacers might well have such deviant intrinsic frequencies that they could not properly be synchronized by the feedback available. In some forms of mutual entrainment (Barlow 1961) the entrained set of oscillators are those with independent frequencies centered about the population average. With the form of interaction assumed here, however, the entrained set consists only of longer-period members of the ensemble. Data demonstrating this are presented in Figure 5.4; in the upper graphs (Parts A and B), the realized average period of a pacer, during interaction, is plotted against the average intrinsic period of the same pacer in the basic (i.e., independent) state; the histograms (Parts C and D) show the overall distribution of realized period values and the basic period values of the pacers involved in these same simulations. The central point illustrated by these graphs is that pacers with intrinsic period longer than some critical value are all engulfed by system performance, all mutually entrained to the same average frequency. Pacers with less-than-critical intrinsic period are on the average somewhat accelerated by system feedback, but not entrained.

This aspect of system behavior is a critical characteristic of the entire class of models, a feature which leads to much of their interesting behavior. It is not simply a peculiar property of the parameter values assumed here, but instead follows inescapably from the nature of the coupling postulated. Brief consideration will show that, in principle, the entrained pacers ought to be all those which have basic periods in the range between the average period of the discriminator cycles, $\bar{\tau}_d$, and $\bar{\tau}_d + \epsilon (1 + \delta_i)$. A pacer with basic period shorter than $\bar{\tau}_d$ cannot be slowed by discriminator activity. A pacer with basic period longer than $\bar{\tau}_d + \epsilon (1 + \delta_i)$ cannot be accelerated sufficiently by the available feedback to remain fully entrained. As can be seen in Figure 5.4, this expectation is qualitatively fulfilled, but stochastic variation in the system means that the boundaries on the range of entrainable basic periods are not sharply defined.

Because the mode of interaction does not permit a pacer to be retarded by feedback, some 50% of the pacers in the ensemble had average periods shorter than that of the discriminator cycles in both of the simulations which led to Figure 5.4. One might be inclined to consider these pacers unimportant, as noise in the system, since they were not synchronized. In fact, however, their behavior is both interesting and important. Their average periods were appreciably shortened by the discriminator, as shown in Figures 5.4A and 5.4B. Of greater significance, however, is the fact that the unentrained pacers showed a phenomenon known as relative coordination

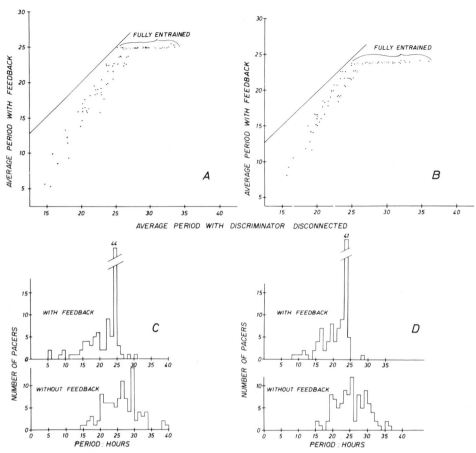

Fig. 5.4 A–D. The effects of mutual entrainment upon period values of individual pacers. Results are illustrated for two simulations, each of 500 h duration, using parameter values of the type model, with 100 pacers in the ensemble. **A** and **B** relationship between average period of a pacer, realized during full-system simulation, and the average period to be expected of that pacer in the uncoupled, basic state [the latter calculated as $\overline{X}_{b,i}(1. + \delta_i)$]. The ensemble output had an average period of about 25 h in case A, and about 24.7 h in case B, corresponding to the asymptotes in values for average period with feedback. This difference in average period between the two simulations was due primarily to differences in anatomy of the two pacemaker ensembles, arising from different arrays of values for $z_{i,\beta}$ and $z_{i,\gamma}$. **C** and **D** histograms of expected average cycle lengths of the pacers in the basic state (without feedback) and of realized average cycle length (with feedback) from the two simulations of parts **A** and **B**, respectively; one-hour bin-width used for grouping data

(von Holst 1939): their discharges scanned various phases of the discriminator output in a nonrandom manner. Figure 5.5A summarizes the data on phase at which the unentrained pacers began to discharge in the simulation of Figure 5.4A. This distribution of phases is a reflection of relative coordination by the unentrained pacers, which can be illustrated schematical-

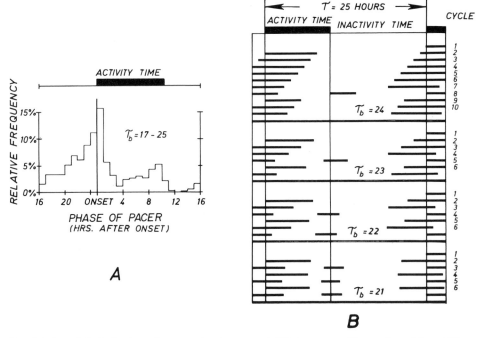

Fig. 5.5 A and B. Behavior of unentrained pacers. **A** Histogram, with one-hour bin width, showing realized phase of onset of discharge for all pacers which had average basic period (without feedback, as defined in the legend of Fig. 5.4) which was less than the average period of the pacemaker, as observed in the simulation which led to Figure 5.4A. Each of these 50 pacers contributed at least 20 values of phase to the histogram. **B** Schematic illustration of expected phasing of discharge (*solid bars*) in successive cycles for pacers with several different values of basic period (τ_b) shorter than that of the system. Cycle-to-cycle stochastic variability set to zero. In a qualitative sense, Figure 5.5A represents a weighted summing of cycle-by-cycle results like those of Figure 5.5B. See Appendix 5.A for full explanation of how the data of Figure 5.5B were calculated

ly as shown in Figure 5.5B. The details of that process are somewhat complex, and are considered in extenso in Appendix 5.A. A full understanding of those details, which can account satisfactorily for all features of Figure 5.5A, is not essential for present purposes. It should be noted, however, that the unentrained pacers often began to discharge within a few hours preceding onset of feedback (the left portion of Fig. 5.5A), thereby contributing significantly to the attainment of threshold during the simulation. Their behavior thus added meaningfully and predictably to stability in timing of the onset of discriminator feedback. Appropriate phasing for that contribution was realized in only a fraction of the cycles of the unentrained pacers, but Figure 5.5A demonstrates that it was a nonnegligible aspect of their behavior.

Data on onsets of discharge by the longer-period pacers, most of which were completely entrained, are summarized in Figure 5.6, again for the sim-

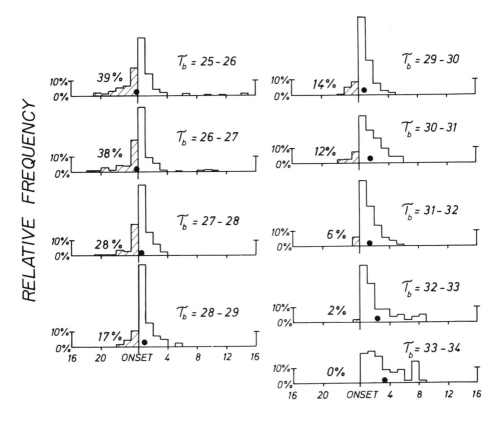

PHASE OF PACER (HRS. AFTER ONSET)

Fig. 5.6. Behavior of entrained pacers. The histograms show realized phase of onset of discharge for pacers which had T_b values (average cycle length without feedback) longer than that of the system, as observed in the simulation which led to Figure 5.4A. *Large dots* within the histograms are positioned at average phase from the entire histogram. Onsets of discharge shortly before onset of feedback are *crosshatched,* with percentage values corresponding to the crosshatched areas. Note that these percentages decrease, and average phase becomes later, as basic period of the pacer becomes longer

ulation which led to Figure 5.4A. The percentage values next to the histograms refer to the fraction of all discharges which began shortly before onset of feedback; they serve to emphasize that the entrained pacers with basic periods close to that of the discriminator cycles were most likely to contribute, in any given cycle, to threshold crossing. Pacers with very long basic periods were largely parasitic on the system: synchronized by the feedback provided, but rarely themselves contributing any share to the pre-onset buildup to threshold. The histograms of Figure 5.6 include a shoulder to the right, which extends several hours after onset of system activity. This distribution of phases of the entrained pacers plays an important role in certain subsequent considerations. It arises because of stochastic cycle-to-cycle variability, quantified by α; feedback from the discriminator does not affect the pacers with an absolute imperative, "Now you must!", as in a deterministic model, but with a persuasive, "Come on! Now is a suitable time!" Their response, with variable latency, reflects a probability-density function describing events which are considered to be intrinsically unpredictable.

The diagrams of Figures 5.4, 5.5 and 5.6 illustrate internal details of system performance which are completely lost in the summary data of Figure 5.3. A plot like Figure 5.3 is sufficient to demonstrate clearly that the system functions adequately: some form of mutual synchronization has resulted which makes summed pacer discharge appreciably more coherent and reproducible than the performance of any single component in the system. Only through examination of the details of single-pacer performance, however, can one fully appreciate how this has been achieved. In spite of strong interaction among pacers, the system evolves a very loose mode of *partial* mutual entrainment. Any summary description of such interactions in terms of either strong coupling or weak coupling would be an oversimplification. A large fraction of the pacers in the ensemble remains unentrained; most of these components of the ensemble have an average period somewhat shorter than that of the driven cycles of the discriminator. The unentrained pacers have a nonrandom distribution of phases, scanning through the pacemaker output in a relatively systematic manner, and thus contributing meaningfully to overall output, even though not truly synchronized. Among the entrained pacers, those with basic periods closest to that of the discriminator contribute most often to attainment of threshold but cycle-to-cycle unpredictability remains a dominant characteristic in their performance, as well as in that of the unentrained pacers.

In subsequent chapters, the output of the pacemaker ensemble will usually be summarized in a format which conveys even less detail than that in Figure 5.3. Only very occasionally will we be referring back to phenomena which are related to those summarized in Figures 5.4, 5.5 and 5.6. It is important to bear in mind, however, that this kind of complexity is inherent in those events which underlie the system performance in all simulations; it is the fine structure from which the global performance of the pacemaker arises.

Appendix 5.A. Relative Coordination of Unentrained Pacers

The behavior of the individual, unentrained pacers, as illustrated in Figure 5.5B, is what is known as relative coordination (von Holst 1939). Those results appear somewhat complex, but a detailed consideration of the first case (τ_b = 24 h, with $\overline{X_b}$ = 16 h, $\overline{Y_b}$ = 8 h) will show that these results were calculated by following very elementary rules. With the Stochastic Coefficient set equal to zero, the calculations do not require a computer. During the first 7 cycles in the upper portion of Figure 5.5B, the pacer functions completely in its basic state, since discriminator activity does not approach the time of onset of discharge; these cycles therefore had periods of 24 h (τ_b), advancing through the cycles in discriminator feedback by 1 h per cycle. In the eighth cycle, the onset of pacer discharge would, without feedback, have occurred exactly 8 h after end of the preceding bout of discriminator activity, but with ϵ of 8 h, the pacer is fully accelerated, and begins its new discharge only 16 h after the last onset of discharge (preceding discharge of 8 h, followed by a recovery phase of 8 h, i.e., $\overline{X_b} - \epsilon$). In this cycle, the pacer discharges for 4 h ($\overline{\delta}$ = 0.5), and then recovers for 11 h, up to the time at which the discriminator becomes active and immediately terminates that phase (a partially accelerated cycle, using only 5 h of the 8 possible). After 5.5 h of discharge in this cycle ($\overline{\delta}$ = 0.5), three basic cycles ensue, with full, normal duration of recovery and discharge. Completely comparable calculations underlie the other cases shown in Figure 5.5B.

The slight peaking in distribution of phases in Figure 5.5A, at the end of discriminator activity, will assume importance in later considerations. It arises because the discriminator strongly accelerates any pacers which might otherwise have begun discharge during the first part of the inactive time. When the discriminator is intermittently producing feedback, there is an interval after the end of feedback, which is as long as the duration of feedback itself, during which very few onsets of pacer discharge are likely to occur. Feedback "reaches out" into this portion of the pacemaker cycle, and consolidates those potential onsets of pacer discharge into a small and rather broad peak, at the end of discriminator activity, as described above in cycle 8.

In Figure 5.5A, there is also a peak, of greater height, immediately after onset of discriminator activity. This arises because an unentrained, short-period pacer, which begins its discharge near the end of discriminator activity, will have a high probability of beginning its next discharge shortly after the next onset of discriminator activity. (An example of this is the second onset of discharge, described above, during cycle 8 in the upper part of Fig. 5.5B.) This is to be expected not only for unentrained pacers with periods very near that of discriminator output, but for a majority of the unentrained short-period pacers (see other portions of Fig. 5.5B). Hence, these pacers, while unentrained, will tend to reinforce the chain reaction which arises as soon as system activity attains threshold, producing a more rapid increase in the number of discharging pacers than would the entrained set alone.

Another conspicuous feature of Figure 5.5A is the broad shoulder of the near-onset peak, sloping off into the hours preceding onset of feedback. The examples of Figure 5.5B indicate the source of this feature: if an unentrained pacer has a basic period only slightly shorter than that of the discriminator, its onset of discharge can be expected to scan the latter portion of the inactive phase of the discriminator cycle (cycles 1 to 7, with $\tau_b = 24$; cycles 1 to 4 with $\tau_b = 23$). Thus, discharges of these unentrained pacers can contribute significantly to the attainment of threshold, and their contribution is not simply noise. It will, on the average, be a relatively predictable component of the gradual rise of summed pacer discharge toward threshold.

Chapter 6
Precision of Model Pacemakers

The focus of attention in this chapter will be upon *precision* of a pacemaker system. We begin with the type model defined in Chapter 5. This will be followed by consideration of many variants, including both modifications of the type model, achieved by changing the values of its parameters, and more extreme alterations, involving new sorts of parameters entirely. An examination of how and why these variations influence precision of the pacemaker system proves to be a convenient vehicle for exposing further details about internal functioning of models of this general class. It serves, in addition, to demonstrate that the type model is by no means an atypical example of its class. Comparable functioning of the models arises even with a wide range of parameter values and formulations, and no trick or "fine tuning" was involved in defining the type model. Finally, this chapter also provides the kind of background which guided me in the choice of parameter values for specific models, to be described subsequently, where the objective is to mimic particular details of the wake-sleep cycle observed of a given species of animal. The general objective here, then, is a further understanding of the pacemaker models, purely as model systems.

Many readers may find such material unnecessary for their interests. None of the issues considered here arises directly from experimental data on the wake-sleep cycle, nor is the material here absolutely essential to the understanding of subsequent chapters. Those who are willing to accept certain assertions without a full appreciation of *why* the system operates in this way may well prefer to skip over the details, and read only the chapter summary, starting on page 79. Beyond that summary, only one issue here will be essential, which is the following: in order to examine temporal precision of the pacemaker system, an identifiable marker point within a cycle of the ensemble behavior is required, some sort of phase-reference point. The two obvious choices available are the onset and the end of feedback by the discriminator, since these are points of discontinuity in the process of mutual entrainment. For reasons which will later become obvious, primary emphasis will be placed here on *onset* of discriminator activity as a phase-reference point.

6.1 Precision, as Affected by Number of Pacers

As one might expect, the number of pacers in the system is a major determinant of precision in pacemaker behavior. Using the format of actograms, with duration of suprathreshold input to the discriminator indicated by solid bars, Figure 6.1 presents examples of the performance of a single pacer (part A), a system incorporating 100 pacers (part B), and a system with 800 pacers (part C). The standard deviation of cycle length, measured between successive onsets of feedback from the discriminator, is a convenient

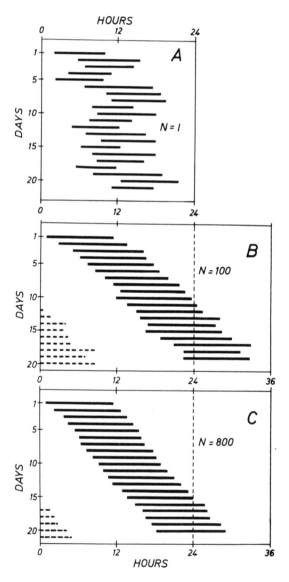

Fig. 6.1. Actograms from simulations with the type model, which incorporated 1, 100, and 800 pacers. Successive 24-h portions of the data have been placed beneath each other, as in the experimental records shown in Figures 2.1 to 2.4. In the upper actogram, the *solid bar* corresponds to discharge time of the single pacer; in the other two graphs, the *solid bars* represent duration of feedback by the discriminator, due to suprathreshold values of summed pacer discharge

measure of system variability; the simulations of Figure 6.1 had standard deviations of 170, 64 and 22 min, respectively.

The standard deviations of onsets from many simulations with the type model are plotted in Figure 6.2, cases in which the only parameter varied was the number of pacers in the ensemble (keeping threshold constant at 0.3N). A line with a slope of − 1/2 appears to describe the trend in the data

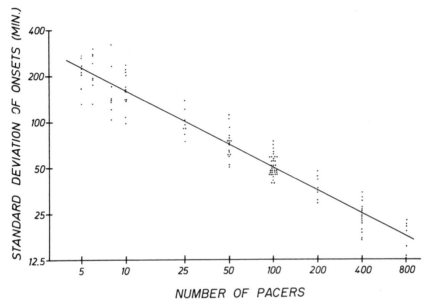

NUMBER OF PACERS

Fig. 6.2. Temporal variability of onsets of feedback, expressed as standard deviation of single-cycle lengths around their mean, as a function of the number of pacers in the ensemble, based on simulations with the type model: double-logarithmic plot. Each *point* corresponds to a separate 500-h simulation, in this diagram as well as in subsequent illustrations. The *fitted line* has a slope of − 1/2, and was drawn through the mean for 100 pacers (50 min)

satisfactorily. Since this is a double-logarithmic plot, that slope implies that standard deviation of cycle length is inversely proportional to the square root of the number of pacers. This empirical result can readily be interpreted in terms of expected values for "waiting times" in a Poisson process. (See App. 6.A for an elaboration on this statement, for which my warm thanks to Dr. John Thorson.)

Since cycle length of a typical pacer in the independent state equals $\overline{X_b}$ $(1. + \overline{\delta})$, the expected standard deviation of single-pacer cycles is about α $(1. + \overline{\delta})$, or 225 min, slightly more than the observed value in Part A of Figure 6.1. The fitted line of Figure 6.2 would not extrapolate back to this value for variability, at N = 1 pacer, but this misfit of the extrapolation should not be disturbing. The behavior of a single pacer acting independently cannot be directly compared with its performance as a member of the ensemble. In the presence of feedback from the discriminator, each pacer alternates between its basic and its accelerated states, thereby greatly increasing the range of cycle lengths which are attainable (cf. Fig. 5.2).

The set of simulations which led to Figure 6.2 demonstrates another property of the models which will be treated later in more detail: the ends of discriminator feedback activity are usually more variable than onsets. Figure 6.3 presents a plot of the standard deviation of cycle lengths, based on

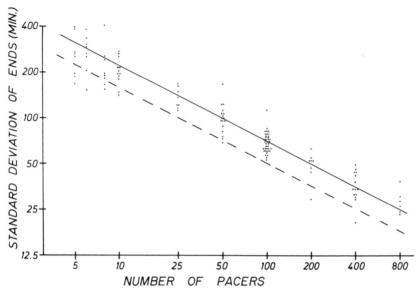

Fig. 6.3. Standard deviation of cycle length, as estimated from ends of discriminator feedback, from the same simulations which led to Figure 6.2, plotted against the number of pacers: double-logarithmic plot. The *solid line* has a slope of $-1/2$, drawn through the mean for 100 pacers (70 min); the *dashed line* represents the fitted line from Figure 6.2, applicable to the data from onsets of feedback. The difference in positions of the two lines means that onset of feedback is considerably more precise than end

ends of feedback (downward crossings of threshold by summed pacer discharge) as a function of the number of pacers. As in Figure 6.2, a line with a slope of $-1/2$ provides an apparently satisfactory fit to the data (solid line), but this line lies appreciably above that which is appropriate for onset of feedback (dashed line); the difference between the two lines indicates that the variability in ends of feedback is about 1.4 times that for onsets of feedback for any value of N.

A qualitative trend corresponding to the difference between these lines (onset more precise than end) might well have been expected from the assumptions, since it is at the onset of feedback that resynchronization of a major fraction of the pacers occurs in each cycle (see Fig. 5.3, 5.5 and 5.6). Even without the variation in delta which arises from γ, the ends of pacer discharge should vary more than onsets, because of stochastic variation in their preceding recovery intervals. It will be shown subsequently that the value assigned to discriminator threshold affects the quantitative relationship between precision of onsets and ends. While onsets of discriminator activity will usually be more precise then ends, the magnitude of the difference shown in Figure 6.3 (a factor of 1.4) is a property of the type model only.

6.2 Influence of Alpha and Beta on Precision

One might well expect that the precision of the pacemaker system, for a given number of pacers, would depend to some extent upon the magnitude of α and β, the coefficients which quantify the cycle-to-cycle and pac-

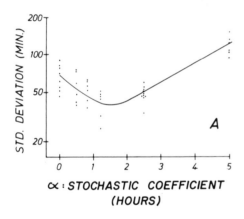

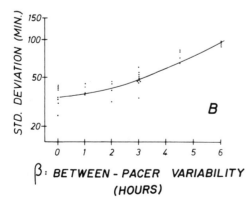

Fig. 6.4 A and B. Standard deviation of onsets of discriminator feedback, as functions of alpha (part **A**) and of beta (part **B**), the coefficients of cycle-to-cycle stochastic variability, and pacer-to-pacer variability, respectively, in recovery interval. 500-h simulations, 100 pacers, with all other parameters as in the type model (including beta for part A, and alpha for part **B**)

er-to-pacer variabilities in recovery interval. Indeed, this is the case; the nature of those dependencies is summarized in Figures 6.4, parts A and B, respectively. The increase in standard deviation due to increased pacer-to-pacer variability (Fig. 6.4B, for which alpha was 2.5 h) seems relatively easy to understand. A small inter-pacer variability will mean that the basic periods of the entrained pacers are all close to (but still predominantly longer than) the period of the system. As shown in Figure 5.6, such pacers contribute to threshold crossing in a large fraction of their cycles by beginning to discharge before onset of feedback. Reducing β therefore increases the number of pacers potentially responsible for threshold crossing, and this will lead to greater precision in a manner comparable, in principle, with the increased precision which results from a larger total population of pacers (Fig. 6.2).

The consequence of altering cycle-to-cycle stochastic variability (Fig. 6.4A) is somewhat more complex, and less in conformity with intuitive expectations; decreasing α increases precision only down to values of α greater than about 1.5 h. (This optimum value can be expected to depend upon the magnitude of β, which was 3 h in the calculations of Fig. 6.4A.) The increase in variability of onsets at smaller α values can, however, be qualitatively understood by considering the limiting case in which α is zero. With no cycle-to-cycle variability, essentially all synchronized pacers will begin to discharge immediately *after* onset of feedback. In contrast to the results shown in Figure 5.6, none of these entrained components would contribute to system threshold crossing: once a pacer has begun discharge before onset of feedback, it has effectively escaped from entrainment. The attainment of threshold must, then, be determined by unsynchronized pacers (cf. Fig. 5.5), and all synchronized pacers are reduced to the status of followers. Thus, when stochastic variability is sufficiently reduced, the number of pacers contributing reliably, in every cycle, to threshold crossing is greatly reduced. This effect on precision can, then, be understood as comparable in origin with the effect of a reduction in N (Fig. 6.2).

6.3 Precision as Affected by ϵ, $\overline{X_b}$ and $\overline{\delta}$

The magnitude of feedback, ϵ, also has some influence on precision of the system (data summarized in Fig. 6.5). As with α and β, the value of ϵ also affects the probability that entrained pacers will begin discharge before threshold crossing, but that influence is indirect. If an entrained pacer, in a given cycle, did not begin to discharge before onset of feedback, then a large value of ϵ will make it very likely that discharge begins immediately after onset of feedback. This will lead to a tighter grouping of phases of the

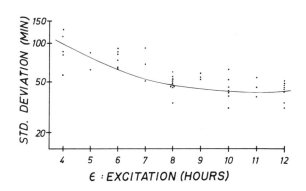

Fig. 6.5. Standard deviation of onsets of feedback, as a function of epsilon, the period-shortening effect of feedback from the discriminator. 500-h simulations, 100 pacers, with all other parameters as in the type model

entrained pacers around onset. In terms of the diagrams of Figure 5.6, the phases immediately after onset will be favored at the expense of the right-hand tails of the distributions. Since, however, a beginning of discharge immediately after onset increases the probability that a pacer might, in the next cycle, begin to discharge *before* onset, the left-hand tails of the phase diagrams in Figure 5.6 will indirectly also be favored, ultimately at the expense of the right-hand tails. The result is that with a larger value of ϵ, a larger fraction of the pacers in the ensemble can contribute meaningfully to threshold crossing: once again, then, a result mediated by an influence on the number of effective pacers.

The overall average charging interval, $\overline{X_b}$, is nothing more than a scaling factor; its value has a direct and apparently linear effect on the period of the system as well as on the duration of feedback by the discriminator. A value of 17 h was assigned to this parameter in the type model in order to produce a period of about 24 h in the ensemble behavior (in conjunction with $\overline{\delta}$ of 0.5); other values of $\overline{X_b}$ between 16 and 18 h are invoked in certain subsequent simulations in which a period somewhat shorter or longer than 24 h was desired in pacemaker performance. When $\overline{X_b}$ was systematically varied from 13 to 22 h, with $\overline{\delta}$ kept at 0.5, there was no detectable influence on system precision.

The Discharge Coefficient, $\overline{\delta}$, which relates duration of pacer discharge to preceding recovery interval, also has a strong effect on period of the system, as is dictated by its role in Eq. (5); this influence can, however, be compensated by an appropriate adjustment of $\overline{X_b}$, so as to leave the period of the pacemaker system near 24 h. With such adjustment, there will of course be a residual effect of $\overline{\delta}$ on the duration of feedback by the discriminator: a proportionately shorter duration of discharge by the individual pacers will lead to shorter duration of suprathreshold input to the discriminator. Furthermore, as can be seen in Figure 6.6, changes in $\overline{\delta}$ (with appropriate adjustment of $\overline{X_b}$, to keep period near 24 h) also have somewhat complex influences on precision of the discriminator cycles.

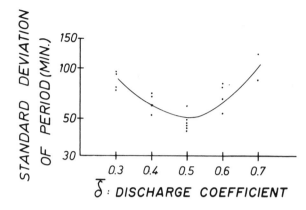

Fig. 6.6. Standard deviation
of onsets of feedback as a function of delta, the average factor of proportionality between
recovery and discharging phases.
$\overline{X}_b$ assigned values for 19.5,
18, 17, 15.5 and 13.5 h, for
δ values of 0.3, 0.4, 0.5, 0.6,
and 0.7, respectively, so as to
keep pacemaker period near
24 h; all other parameters as
in the type model; N = 100
pacers

The curvilinear relationship of Figure 6.6 can be interpreted by examining the effects of $\overline{\delta}$ upon the behavior of the unentrained (short-period)
pacers. When $\overline{\delta}$ assumes a value of 0.3, the duration of feedback due to
discriminator activity is about 7 h in a 24-h cycle. The consequences of
such a short duration of feedback is that the fraction of the cycles of unentrained pacers which can contribute to threshold crossing will be smaller
than when $\overline{\delta}$ has a value of 0.5, because many of their discharge phases,

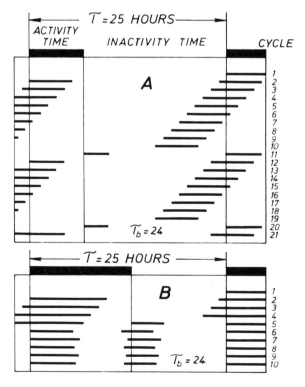

Fig. 6.7 A and B. Schematic
illustration of expected phase
of start of discharge in successive cycles (alpha set to zero)
for a pacer with basic period
of 24 h, when system feedback
activity has a period of 25 h.
A $\overline{\delta}$ = 0.3, with a resultant duration of system feedback of 7 h.
Note that in cycles 7 through
10, and cycles 16 through 19,
the end of discharge occurred
before onset of feedback. **B** $\overline{\delta}$ =
0.7, with resultant duration of
feedback of 13 h. Note the
approach to steady state in
which discharge of this pacer
does not occur during the interval preceding onset of discriminator activity. See text
for further discussion

which start during the time when the discriminator is inactive, will end before threshold is attained (compare Fig. 6.7A with Fig. 5.5B). When $\bar\delta$ assumes a value of 0.7, the duration of feedback is about 13 h in a 24-h cycle, and most of the unentrained pacers will tend to begin discharge either immediately after onset of feedback or shortly before the end of feedback (Fig. 6.7B), thus making no contribution to attainment of threshold. A value of $\bar\delta$ near 0.5 thus proves to be optimal for precision in system behavior, because it permits maximum participation by the unentrained pacers in timing the onset of feedback.

6.4 Threshold of the Discriminator

The threshold of the discriminator specifies the number of simultaneously discharging pacers necessary for feedback to be produced. The value assigned to this parameter has complex effects on the precision of onsets of discriminator activity, as shown in Figure 6.8A. This curvilinear relationship is a consequence of the manner in which threshold influences other aspects of system behavior, as illustrated in Figure 6.9: an increase in discriminator threshold leads to an increase in period of the driven cycles of the discriminator, as well as to a shorter duration of feedback. Both of these latter result are expected consequences of the basic functioning of the system: the "waiting time" for the attainment of a threshold of 40% of the total pacers would necessarily be longer than that required for a threshold of 30%, thereby lengthening the interval between successive onsets of system activity; and unless $\bar\delta$ is considerably greater than 1.0, the downward crossing of a threshold set at 40% will occur sooner after onset of feedback than if threshold is 30%, leading to a shorter duration of system activity in each cycle.

Recalling that the entrained pacers consist of those with basic periods approximately in the range between the period of the discriminator cycles, $\bar\tau_d$, and $\bar\tau_d + \epsilon(1. + \bar\delta)$ (Fig. 5.4), one would expect an increase in the period of the system, caused by an increase in threshold, to decrease the number of entrained pacers (Figs. 6.10A and 6.10B), unless ϵ is very small relative to β. This reduction will, in turn, decrease precision at high threshold, an effect comparable with a reduction in the total number of pacers (cf. Fig. 6.2).

When, however, threshold is decreased to the extent that the period of the system becomes appreciably shorter than the average basic period of the general ensemble of pacers (Fig. 6.10C), a conflicting process arises, associated with the assumed Gaussian distribution of recovery intervals. Entrained pacers with basic periods only slightly longer than the period of the

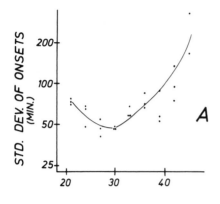

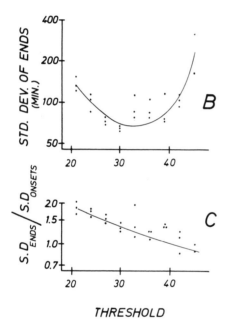

Fig. 6.8 A–C. Variability of pacemaker cycles, as a function of θ (threshold, expressed as the number of pacers required for feedback to occur). **A** standard deviation of onsets of feedback; **B** standard deviation of ends of feedback; **C** ratio of standard deviations of ends of feedback to standard deviation of onsets. Based on 500-h simulations with 100 pacers, with all other parameters in the type model

system are those most likely to contribute to the attainment of threshold, by beginning to discharge somewhat before onset of feedback in a significant fraction of their cycles (compare percentage values in Fig. 5.6). If, due to lowered threshold, the period of the system becomes appreciably shorter than the modal value for basic period of the ensemble, then a smaller fraction of the entrained pacers will have basic periods appropriate for maximum contribution to achievement of threshold (Fig. 6.10C).

Another factor, which would also tend to produce maximum precision at intermediate values of threshold, is associated with the influence of threshold on duration of feedback, which, in turn, affects phasing of the unentrained pacers. As shown in Figure 6.7, a duration of feedback as short

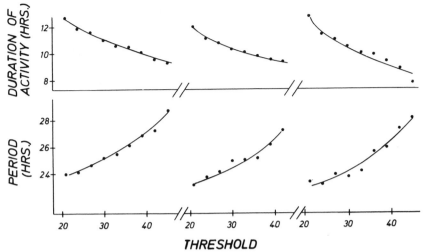

Fig. 6.9. Duration of feedback activity, and average period length, as functions of threshold, for three series of simulations, each with a different set of pacers using type-model parameters (different arrays of $z_{i,\beta}$ and $z_{i,\gamma}$). Small but consistent differences in behavior among the three sets of simulations are evident, which arise due to the fixed properties of the pacers within the ensemble, as dictated by the arrays of values for $z_{i,\beta}$ and $z_{i,\gamma}$; note, for example, that in the second set of data, duration of feedback and average period were slightly shorter, at equivalent threshold values, than in the first and third sets of data. Such differences can be taken as equivalent to differences between pacemakers in their anatomy, arising by chance in the assembly of finite arrays: differences of the sort to be expected between individual animals

as 7 h, or as long as 13 h, tends to reduce the effective contribution of unentrained pacers to attainment of threshold; a larger fraction of their discharge phases do not contribute to onset of feedback. Hence, an intermediate duration of discriminator feedback, associated with an intermediate value of threshold, should also contribute to minimal variability in system performance.

The influence of discriminator threshold upon the variability of the ends of pacemaker activity is summarized in Figure 6.8B; as shown in Figure 6.8C, the variability in ends of feedback, *relative* to the variability in onsets, decreases monotonically as threshold increases. The precision of ends of pacemaker activity will be coupled to a large extent with the precision of onsets, since the end of feedback is primarily determined by the end of discharge of those pacers which begin to discharge immediately after onset of feedback. Thus, it is not surprising that the trends in Figures 6.8A and 6.8B are more or less parallel. The result of Figure 6.8C is, however, more difficult to interpret, since it is the net consequence of several conflicting influences. There are certain ways in which lowering threshold should be expected to increase precision of ends of feedback and other ways in which it ought to decrease this precision, in terms of effects upon

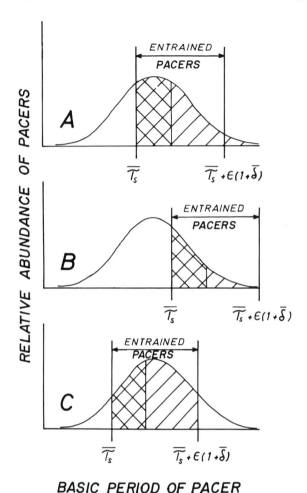

the entrained pacers; and comparably complex and conflicting tendencies
are also involved, mediated by the influence of threshold on the duration
of feedback, as this factor determines the phasing of unentrained pacers.
Since it is not obvious which of these processes would dominate in determining the trend shown in Figure 6.8C, the result there is probably best
left as an empirically established property of the system.

6.5 The Implications of Experimentally Observed Precision

If the parameter values, which were so arbitrarily assigned to the type
model, were by coincidence to correspond to biological reality, an extrapolation of the relationship between standard deviation and number of pacers

would suggest that some 28,000 pacers would be necessary in the ensemble, in order to account for the precision observed in the best of circadian records (Figs. 2.1 and 2.3, with standard deviations of about 3 min). The values assigned to α, β, and γ in the type model, however, were deliberately chosen as extreme assumptions, so as to test the robustness of the class of models. From a more optimistic viewpoint, one might propose to reduce the values of these parameters appreciably; the pacers of the ensemble might well be less sloppy than proposed in the type model. How much they might plausibly be reduced is unclear. In my own opinion, it does not seem unreasonable to propose that α be reduced to half its value in the type model, and β, and γ to one third the type-model values (i.e., 1.25 h, 1 h and 1/24 for α, β, and γ, respectively)[2]. A set of ten simulations incorporating these values, with N = 100 pacers and with ϵ, $\bar{\delta}$, and θ as in the type model, led to a mean estimate for standard deviation of onsets of 17 min (range 12 to 21 min in ten simulations). If we take the best of these estimates (to compare with the best of experimental data, like those of Figs. 2.1 and 2.3) and extrapolate to a larger ensemble on the basis of proportionality to $\sqrt{1/N}$, the precision comparable with the best of circadian records might be achieved with an ensemble of about 1600 pacers.

No reliance can, however, be attached to this value. The calculation is based on my own guesses about plausible values for α, β, and γ; beyond this, it may be that other, uninvestigated combinations of parameter values (including ϵ, $\bar{\delta}$, and θ) might produce greater precision, because of interdependence in their effects. Moreover, other factors which affect precision without increasing the number of pacers or reducing the values of α, β, and γ, are described below as well as in Chap. 8.

A practical limitation in further exploration of this issue is that in simulations like these, in which standard deviation becomes as small as 10 to 20 min, extraneous variability is introduced into the estimate of precision because of the discrete-time nature of the calculations, with an ATU of 15 min. We are left, then, with the relatively unsatisfying conclusion, which is largely a matter of subjective interpretation, that a pacemaker with not more than a few thousand pacers might well be able to account for the maximum precision observed in experimental situations.

---

2 Given these parameters and the type-model values for X_b and δ, the standard deviation of inter-pacer variability for average total cycle length of a pacer in the basic state will be 1.66 h [i.e., $\sqrt{X_b^2 \gamma^2 + (1 + \delta)^2 \beta^2}$], and the average intra-pacer stochastic variability will be 1.88 h [i.e., $\alpha (1 + \delta)$]. Comparable values for intercellular and intracellular circadian variability of *Acetabularia* (Karakashian and Schweigert 1976) can be estimated at 1.92 and 2.1 h, respectively (see Chap. 5).

6.6 Alternative Formulations of Parameters

As was noted earlier, the formulation of the duration of pacer discharge as a proportion of the preceding recovery interval [Eq. (3)], as well as the formulation of the effect of feedback as a constant value for shortening the recovery interval, represent rather arbitrary choices; such formulations are compatible with the expected behavior of a simple relaxation oscillator, but they are by no means essential characteristics of the general class of models considered here. Additional computer simulations have demonstrated, however, that the performance as well as the precision of the pacemaker system are not appreciably changed by altering the formulations used for discharge and for feedback. The alternatives considered here are, of course, by no means an exhaustive coverage, but only a sampling of conceivable modest modifications.

The duration of pacer discharge was specified in Chap. 4 as a proportion, $\bar{\delta}$, of the preceding recovery phase [the Discharge Coefficient of Eq. (3)], which was subject to a pacer-to-pacer variability quantified by γ [Eq. (4)]. Precision of the pacemaker system is only slightly improved, relative to the type model, when γ is set to zero, resulting in fixed proportionality between durations of recovery and discharge which is invariant among pacers. (Compare columns A and B of Fig. 6.11). Slight further improvement in preci-

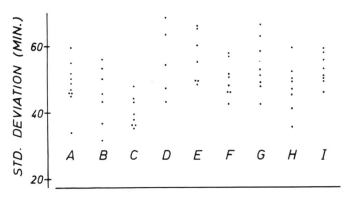

Fig. 6.11. Standard deviation of onsets of feedback with alternative formulations for various parameters. *A* type model; *B, C,* and *D* alternative formulations for duration of recovery; *E, F,* and *G* alternative formulations for period-shortening effect of discriminator; *H* and *I* inter-pacer variability incorporated into effects of pacer on discriminator. 500-h simulations, 100 pacers. See text for particular formulas used for each data column

sion results (column C of Fig. 6.11) if the discharge phase of each pacer is made identical and independent of the preceding recovery phase ($Y_{i,j} = K$, with K set to 8 h, so as to produce a circadian period in ensemble output).

Neither of these alternative formulations is appealing, however, because they unrealistically propose complete identity among pacers in a critical aspect of their performance. To reinject pacer-to-pacer variability in properties, and nevertheless dispense with proportionality between durations of recovery and discharge, the discharge phase can be specified as $Y_{i,j} = K + \lambda z_{i,\lambda}$, in which K is the ensemble average duration of discharge, $z_{i,\lambda}$ is a pacer-specific unit normal variate, and λ is the standard deviation of inter-pacer variability in hours. The pacemaker functions quite normally with this sort of formulation; and when λ is assigned a value of 2 h (with K again being 8 h), precision in ensemble output is quite comparable with that of the type model (column D in Fig. 6.11). The only aspects of realism missing here and present in the type model are cycle-to-cycle stochastic variation in duration of discharge, and inter-pacer variability in the effects of stochastic, cycle-to-cycle variability.

Similarly, the functioning and precision in system behavior were not affected appreciably by several alternative formulations for the magnitude of the effect of discriminator feedback. Column E in Figure 6.11 presents data on standard deviation of onsets from simulations in which pacer-to-pacer variability was included in the effect of feedback. This was accomplished by using the equation, $\overline{X}_{a,i} = \overline{X}_{b,i} - \epsilon + \eta z_{i,\eta}$ as an alternative to Eq. (6), where $z_{i,\eta}$ is a unit random normal variate, with ϵ assigned a value of 8 h, as in the type model, and η a value of 2 h. For column F, the difference between recovery interval of a pacer in the basic and accelerated states was formulated so that the effect of feedback would be proportional to average recovery interval ($\overline{X}_a = K' \overline{X}_b$, in which K' was assigned a value of 0.5). For the calculations of column G in Figure 6.11, a similar proportionality for the effect of feedback was used, with the additional inclusion of pacer-to-pacer variability in the proportion [i.e., $\overline{X}_{a,i} = (K' + \nu z_{i,\nu}) \overline{X}_{b,i}$, where $z_{i,\nu}$ is a unit random variate, with K' set at 0.5 and ν at 0.1].

To this point, all simulations assumed that each pacer, during its discharge, contributed identical input to the discriminator. No major change in the system performance or precision is introduced, however, by the assymption of pacer-to-pacer variability in this input. For the calculations leading to columns H and I in Figure 6.11, the contribution of the i th pacer to the summed input was expressed as $1.0 + \mu z_{i,\mu}$, where $z_{i,\mu}$ is a unit random normal variate, with μ assigned a value of 0.1 for column H and 0.2 for column I.

There are differences which are statistically significant among the data in the various columns of Figure 6.11, but the means are all within about 20% of that of the type model, and the basic performance of the system was apparently unchanged. Furthermore, as is shown in Figure 6.12, dependence of system precision upon the number of pacers appears to be equivalent to that of the type model in each of these formulations (i.e.,

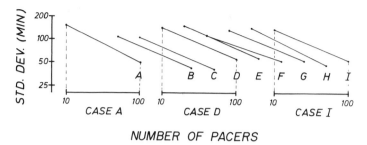

NUMBER OF PACERS

Fig. 6.12. Effect of varying pacer number upon standard deviation of onsets of feedback for the nine formulations shown in Figure 6.11: double-logarithmic plot, with new abscissa values for each data set, examples of which are shown for cases *A, D,* and *I.* The *lower point* for each line is the mean of the data presented in Figure 6.11; the *upper point* is the mean based on six or more simulations with ten pacers. Note that the slopes of the lines are generally comparable with that of the type model (case *A*), i.e., $-1/2$

standard deviation inversely proportional to the square root of the number of pacers in the ensemble). That result is apparently quite general for systems of this sort. (See App. 6.A for a consideration of why this should be so.) Thus it appears that we have not, in the alternatives considered here, tampered with any critical characteristic of the system; and it appears, furthermore, on the basis of the square-root relationship, that even the most extreme of the variants considered in Figure 6.11 could be made equivalent in precision to the type model, simply by adding or subtracting some 40% to the number of pacers in the system.

6.7 The Relationship Between Pacer Input and Discriminator Output

In all the simulations considered to this point, the discriminator has maintained its properties as an on-off switching element: as long as the summed discharge of all pacers exceeds a fixed threshold, the discriminator remains fully active. A consideration of other, less discontinuous modes by which pacer discharge might be transformed to discriminator output will be postponed to Chap. 8; this is an important issue, which deserves special attention, but that will require material presented in Chap. 7. It is useful at this point, however, to note that simple arithmetic summing of the pacer discharge for comparison with discriminator threshold, as specified in Chap. 4, is a calculational procedure which need not imply linearity in the physiological mechanisms involved. The manner in which the pacers affect the discriminator at a mechanistic level could be strikingly nonlinear without altering system performance at all. The requirement is only that the effect of pacers on the discriminator be a monotonic function which increases with the number of discharging pacers; and that onset of discriminator activity arise at some fixed value of this function, which fixed value then corresponds to threshold. For example, if change, say, in membrane potential of the discriminator might be better described as $(N_D)^x$, where N_D is the number of discharging pacers and x is any positive number, then equivalent functioning of the models is obtained simply by specifying that the discriminator produces feedback (ϵ hours) whenever pacer-induced change in membrane potential exceeds $(\theta)^x$. The determination of "summed pacer discharge" in the simulations is therefore without implications about functional detail.

6.8 Stochastic Variation in Threshold[3]

The mechanism underlying the dependence of system precision upon the number of pacers in the ensemble can be illustrated schematically for the type model as shown in Figure 6.13. Brief examination of this diagram will reveal an oversimplification which has been tacitly ignored in assigning a fixed value to threshold: no allowance was made for possible stochastic variation in the parameter θ, as represented by PD_θ in the upper illustration. If a noise component which is large relative to stochastic variation in the number of pacers discharging (PD_N) were to be associated with threshold, then that noise would be the dominant factor in determining system precision, and the number of pacers in the ensemble would become irrelevant. In more quantitative terms, assume that noise in discriminator threshold is independent of pacer activity; then the standard deviation of system performance, s_O, should be approximately proportional to $\sqrt{s_P^2 + s_\theta^2}$, where s_P^2 is the variance associated with PD_N in Figure 6.13, and s_θ^2 (which has until now been neglected) is the variance associated with PD_θ. It is unrealistic to assume for a biological system that discriminator threshold has negligible temporal variability, and no matter how small s_θ^2 is assumed to be, that factor must set some upper limit on the pacemaker precision which is attainable by increasing number of pacers in the ensemble.

A full evaluation of this asymptote by means of computer simulations would require a two-way array of calculations, in which the number of pacers and the magnitude of noise in threshold were jointly varied; such simulations have not been undertaken. It is possible, however, to arrive at a rough estimate of the relative significance of s_P and s_θ in the type model from considerations based on Figure 6.13. Let us assume, for example, that $s_\theta = 3\%$ of the value of θ; with threshold being 0.3N in the type model, threshold then varies stochastically with a standard deviation of about

---

3 My thanks to John Thorson for collaboration on this analytical approach.

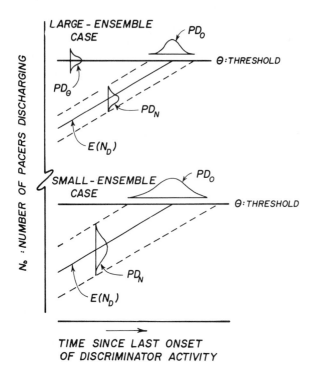

Fig. 6.13. Comparison of probability-density functions for large-ensemble and small-ensemble pacemakers. $E(N_D)$ is the expected many-cycle average for the behavior of the number of pacers discharging; PD_N is the probability-density function for cycle-to-cycle variability in number of pacers discharging, with dashed lines to indicate that this distribution applies throughout the increase in $E(N_D)$; PD_O is the resulting probability-density function for the time of onset of discriminator activity; PD_θ is the probability-density function associated with temporal variation in the value of threshold, which is neglected in the type model. With a large ensemble (*upper diagram*), the stochastic variance of PD_N will be smaller than with a small ensemble; if PD_θ is ignored, this decrease in PD_N leads directly to greater precision in system output

0.01N. To estimate the efficacy of such variation, let us now determine the size of the ensemble for which the expected value of s_P would be just equal to this value of s_θ. Referring to Figure 5.3, note that the number of discharging pacers increases approximately linearly between hours 360 and 365, at a rate of about 40 pacers (i.e., 0.05N) per hour; this value can be used as the slope of the line $E(N_D)$ in Figure 6.13 where it is evident that this slope represents the coefficient of proportionality between PD_N and PD_O, and hence, between s_P and s_O (the standard deviation in number of discharging pacers and that in the temporal distribution of threshold crossings, respectively); that is, $s_P = 0.05 s_O$. If s_P is to be 0.01N, then $s_O = 1/5$ h or 12 min. In Figure 6.2, we have established that $s_O = k\sqrt{1/N}$, with k being approximately 500 min. From that relationship, we see that an ensemble of about 1700 pacers would have an s_O of about 12 min. For small-er values of N, therefore, 3% fluctuations in threshold will be relatively in-efficacious, since s_θ will be less than s_P. From the nature of these calcula-tions, it is also evident, for example, that if stochastic fluctuations in thresh-old had twice as large an amplitude, that variability would become the dom-inant component determining system precision for ensembles larger than about 1700/4 or 425 pacers.

It should be noted that the frequency characteristics (spectral distribu-tion) of the noise have not been taken into account in these calculations. It is clear, however, that we have only been dealing with fluctuations which have frequency low enough so that threshold does not change greatly dur-ing the transit of $E(N_D)$ through the region determined by PD_O and PD_N (frequencies corresponding to changes over hours or days). Higher-frequency noise would in fact tend to reduce the variance in period associated with that sort of threshold variation; and high-frequency noise would have to be appreciably larger in amplitude in order to be the equivalent of the low-frequency noise to which these calculations apply.

6.9 Summary of Factors Influencing Precision

Simulations in which the number of pacers in the type model was varied demonstrate that the standard deviation of period, between successive on-sets of feedback, is inversely proportional to the square-root of the number of pacers present in the system; i.e., $S = k\sqrt{1/N}$, where S is the standard deviation in minutes, and N is the number of pacers. For the type model, $k \approx 500$ min. Ends of feedback are more variable than onsets, but the square square-root relationship with N still holds; k is about 700 min for ends of feedback for the type model.

The coefficients β and α, which quantify inter-pacer and intra-pacer vari-ability respectively, are by no means equivalent in their effects on system performance. Increases in β lead to monotonic increases in the standard deviation of system performance; the minimum value of k (see above) is about 350 min, at β equal to zero. The relationship between α and system precision is not monotonic. At optimum precision (all other parameters of the type model held constant), α has a value of 1–2 h; in this range, k has a value of about 400 min. Increases in ϵ, the magnitude of feedback, im-prove precision somewhat. The relationship appears to be monotonic, at least up to a value of 12 h for ϵ, but the improvement over the type-model value (8 h) is minor (k equal to about 450 min at ϵ = 12 h). $\overline{X}_b$, the overall mean recovery interval of the ensemble, is a scaling factor, and has no de-tectable effect upon precision, when varied between 13 and 22 h, although, of course, the ratio of variability to mean period decreases with increase in $\overline{X}_b$.

Delta, the coefficient relating recovery to discharging intervals, directly affects period of the system also; when these effects are compensated by appropriate changes in $\overline{X}_b$, the relationship between precision of the system and δ is not monotonic; optimum performance is associated with values near that of the type model, i.e., with δ near 0.5. Variations in the threshold of the discriminator also show a nonmonotonic relationship with precision, in addition to their monotonic effects upon period length of the system and on duration of feedback activity by the discriminator. Optimum precision, with all other parameters as in the type model, is attained with θ in the range between 0.25 N and 0.35N, that is, near the type-model value of 0.3N.

All these effects of varying α, β, δ, ϵ and θ can be interpreted as being mediated by influences on the size of that subpopulation of pacers which contributes to attainment of threshold: the more pacers participating, the greater the precision. Simulations based on choices for α, β and γ, which do not appear to me to be unduly optimistic assumptions, lead to the suggestion that a pacemaker ensemble of some 1600 pacers of the type considered might be sufficient to account for the kind of precision observed in the best of circadian experimental records. More demanding assumptions, involving reduced variability between pacers and greater intrinsic reliability, as well as certain considerations described subsequently (Chap. 8) would permit this estimate to be revised downward. On the other hand, possible stochastic variation in threshold of the discriminator has been neglected in all these calculations, and that factor alone will asymptotically set an upper limit on precision of pacemaker output regardless of the number of pacers in the ensemble. The size of the ensemble for which stochastic variation in threshold will become important depends upon the magnitude of that variation and its frequency spectrum, as well as upon many other parameters of the model. Using type-model parameters, it appears that the low-frequency, day-to-day variation in threshold can be on the order of 3% and still be unimportant for system precision when the size of the ensemble is less than about 1700 pacers.

A number of alternative formulations for the duration of discharge by a pacer, and for the effect of discriminator feedback, have been investigated, as well as the incorporation of variability between pacers in their effects upon the discriminator. None of these modifications, which involved adding or deleting parameters or changing their nature, conspicuously altered the basic behavior of the system; and, within the range of parameter values considered, effects on precision of the system were small. Standard deviation of cycle length was altered by no more than about 20%. Since the inverse square-root relationship between standard deviation and number of pacers present seems to hold throughout all cases considered, any change in system precision induced by modest reformulation of a model of this general class could apparently be compensated by appropriate modest increases or decreases in the number of pacers in the ensemble.

Appendix 6.A. On the Influence of N upon Precision
(by John Thorson)

Although the system is designed for simulation rather than for analysis in terms of the expected behavior of particular random variables, the observed proportionality between standard deviation and $N^{-1/2}$ can readily be given a plausible analytical basis. Take for example the duration of the subinterval defined by the nearly linear rise of the fraction discharging prior to activity by the discriminator (see Fig. 5.3). This subinterval is terminated by the discriminator sensing the $0.3N^{th}$ event in a series of random events (initiations of discharge) occurring at nearly a constant rate (i.e., constant probability per unit time) in this region. Since this rate is itself proportional to the number of pacers in the system, the subinterval can be looked upon as the waiting time for the rth event in a Poisson process with rate λ. This random variable (waiting time) is well known (see, e.g., Parzen 1960, p 260) to be gamma distributed with mean r/λ and variance r/λ^2. Since both r and λ are proportional to N, the result that the variance is proportional to $1/N$ follows immediately.

The system is of course orders of magnitude more complicated than in the above idealization, and other more intricate consequences of the law of large numbers certainly apply. Note that the termination of the period of feedback activity is similarly determined by a waiting time until the discharging fraction has fallen to 0.3N. Hence, similar variation will be introduced, and the variance of the total period will be determined by the interaction of several such interdependent phenomena via the strongly nonlinear conditions imposed. It seems noteworthy, though, that the variance of the total period vs N conforms closely to that predicted above from the most elementary considerations. The result is assured, apparently, by the many stochastic effects introduced.

Chapter 7
Influences of Constant Light Intensity

As the preceding chapter has made clear, the interpretive vehicle proposed here, to account for general properties of the wake-sleep rhythm, is not a unique and highly specified model for the pacemaker system, but instead a broad general class of models. In order to avoid extensive repetition in this and subsequent chapters of phrases like "the class of models which is being considered here," or "the general kind of models under examination," it will be convenient to have a shorthand name; for that purpose, I will use the term, Coupled Stochastic System.[4] A Coupled Stochastic System is here defined, then, as a model of that general class which incorporates the kind of mutual coupling described in Chap. 4, although not necessarily based upon an identical set of equations or parameters.

In this chapter we undertake the first of several direct comparisons between performance of Coupled Stochastic Systems and experimental recordings of the locomotor activity of birds and rodents. For that kind of comparison, further specification is required, since up to this point all descriptions of pacemaker performance from simulations have been couched in terms either of summed pacer discharge, or of pacemaker feedback activity: the input and output of the discriminator, both of which are presumed to consist of trains of nerve impulses. To bridge the gap between hypothetical nerve activity and whole-animal behavior, one conceivable approach would be the idea that summed pacer discharge is in some way comparable with degree of the animal's arousal or its level of excitement, but since "level of excitement" has not been experimentally measured, that sort of interpretation would provide no direct basis for comparison between experimental data and performance of the models. The on-off behavior of the discriminator offers a more promising basis for comparison, since it involves two mutually exclusive alternative states of the system, which might be viewed as analogous to wakefulness and sleep. Furthermore, recordings of locomotor activity, which serve here as an index for times of wakefulness, also often show a striking on-off dichotomy. As will be shown, both here and in subsequent chapters, that approach proves fruitful: recorded patterns

---

4 Straitlaced colleagues have dissuaded me from my original designation, the acronym PRECISE, from Pacemaker Resulting from Excitatory Coupling of Independent Stochastic Elements.

of locomotor activity show many detailed resemblances with patterns of discriminator activity. These similarities represent the empirical justification for the assumption proposed here: that times of discriminator feedback ordinarily correspond with times of wakefulness. The interpretation, then, is that when summed pacer discharge exceeds discriminator threshold, the discriminator not only produces that output which constitutes feedback on the ensemble of pacers, but also has a second output which ordinarily provides permissive conditions for locomotor activity.

It is not, of course, essential to assume that the two striking discontinuities in pacemaker behavior correspond directly to the discontinuities in an activity recording. One might, for example, entertain the suggestion that a threshold for locomotor activity and wakefulness be set at some level of summed pacer discharge, which differs from the threshold value above which discriminator feedback affects the pacer ensemble. The model would, however, thereby require an additional parameter, which is an unnecessary luxury unless we should find evidence that such a parameter would contribute to a better description of experimental data.

An alternative possibility, which is equally economical in terms of parameters, is to assume that times of discriminator feedback correspond to times of sleep rather than of wakefulness. That interpretation would, however, conflict strongly with a consistent aspect of the experimental data. The activity records of Figures 2.1 to 2.4 are all cases in which onsets of locomotor activity were conspicuously more regular in their timing than ends of activity. This phenomenon is extremely general in free-running rhythms of birds and rodents: so common, in fact, that it can usually be taken for granted without further comment. The generality of this empirical observation, and the finding that Coupled Stochastic Systems usually show appreciably greater precision in onsets of discriminator feedback than in ends (e.g., Figs. 6.3 and 6.8) together constitute my initial rationale for equating times of discriminator feedback with wakefulness rather than with sleep.

The environmental light regime constitutes the input to the pacemaker, which is of interest here. A number of other kinds of stimuli are known to affect the circadian rhythms of birds and rodents (e.g., temperature; certain chemicals; mechanical stimuli including vibration, noise and atmospheric pressure), but most of those effects tend to be small in magnitude compared with the profound influences of light. Furthermore, a broad spectrum of different experiments have been conducted, based on varying the lighting regime, but data on other sorts of input to the circadian system are few in number, a disparity which arises primarily because it is such a simple matter to manipulate lighting conditions.

In this chapter, the focus will be upon the effects of what can be termed DC input: different intensities of continuous illumination, with consideration of cyclic and pulse inputs postponed to subsequent chapters. In order to proceed further, some decision is necessary about where and how light exerts its influence in a Coupled Stochastic System. The effects of light on biological systems are sufficiently diverse as to permit great freedom in assumptions; light might conceivably affect any one of the parameters of Coupled Stochastic Systems. As shown below, however, experimental data on the wake-sleep cycle under constant conditions of light place major constraints on the decision about how best to introduce effects of light into the pacemaker models.

7.1 Effects of Light upon Free-running Period

One of the central generalizations from studies of the circadian wake-sleep rhythm of vertebrates, and a central issue in this chapter, is the finding that the intensity of constant light which is imposed upon an animal as a continuous stimulus has a systematic effect on the frequency of the observed circadian rhythm. A comparison of this experimental result with the results of simulation with the pacemaker models, such as those described in Chap. 6, serves to demonstrate that five of the parameters in the models are unlikely candidates as mediators for the action of light upon the system. Changes in the values of α, β, γ, ϵ, and N have no appreciable effect upon pacemaker period, and constant light *does* affect period, so these parameters can be tentatively excluded from consideration.

In evaluating the three remaining parameters ($\overline{X}_b$, $\overline{\delta}$, and θ, i.e., Mean Duration of Recovery, Discharge Coefficient, and Threshold, respectively) as candidates, another empirical generalization from data on birds and rodents under constant conditions deserves consideration: those changes in the level of continuous light intensity which lengthen the period of a circadian rhythm also usually shorten the per-cycle duration of locomotor activity. Decreasing or increasing the cycle length by means of light is thus accompanied by an opposite change — sometimes a very large alteration —

Fig. 7.1 A—D. Examples of the effects of the intensity of constant light upon the activity rhythms of birds and rodents. **A** a house finch, with lighting changed at the time indicated, from modest intensity to very dim; **B** a chaffinch, which experienced modest intensity, then dim, and then modest intensity again (from Aschoff et al. 1952). **C** a hamster, with a change from bright continuous light to complete darkness (from Pittendrigh 1967); **D** a deermouse, with change from complete darkness to bright light (ibid.). The actograms of Parts **C** and **D** represent "double plots"; the complete record was duplicated and placed to the right of the original, raised vertically by one day, to preserve continuity. A presentation of this sort facilitates recognition of features of the record which scan across midnight

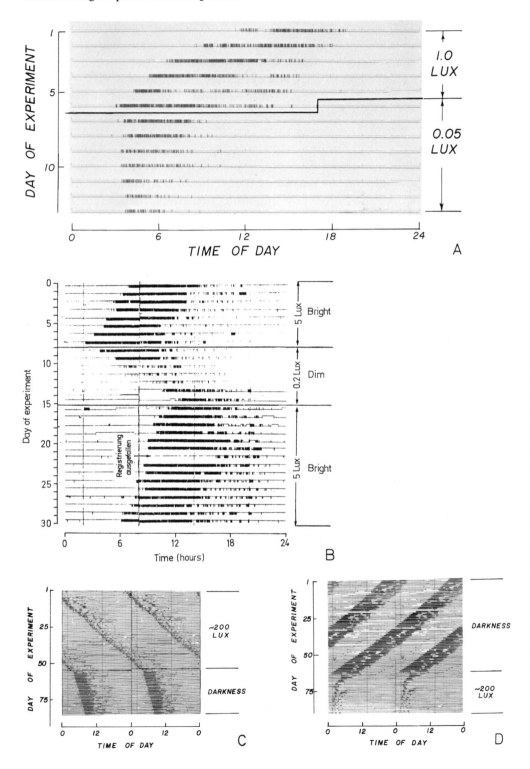

in duration of wakefulness. No simple formulation of the influences of light upon either $\overline{X}_b$ or $\overline{\delta}$ can accomplish this; in order to increase the duration of discriminator feedback, $\overline{\delta}$ must be increased, but this produces an increase in the length of the total cycle, not a shortening.

It is concievable, of course, that light might affect period of the pacemaker by altering $\overline{X}_b$ and $\overline{\delta}$ in opposite directions; with appropriate choice of the relative magnitudes of these two effects, a fit to experimental data on both period and duration of activity could be achieved. Before resorting to this or other comparably complex treatment, however, we should consider, as a candidate, the remaining parameter of the models, θ. As will become evident, both here and in subsequent chapters, the assumption that light affects only threshold of the discriminator proves to have a number of advantages. The first of these is illustrated for the type model in Figure 6.9: an increase in discriminator threshold leads to lengthening of the period of the system and simultaneously induces shortening of the duration of feedback (locomotor activity) as the experimental data demand. Thus, one can assume that light directly alters only one parameter of the model system, and obtain a fit to both observed features of the data.

At this stage, however, the experimental data should be confronted more directly. Four representative examples are shown in actogram format in Figure 7.1. Several different ways of summarizing these kinds of data on the influences of light on free-running period are shown in Figure 7.2. The initially disconcerting aspect of the observations is that there are some species for which light lengthens the period of the wake-sleep rhythm, and others for which it does just the opposite. Clearly, the assumption that light acts in a unique manner upon discriminator threshold cannot account for both kinds of result.

Fig. 7.2 A−F. Examples of the kinds of data which illustrate the effects of constant lighting upon the period of free-running circadian rhythms. A Inter-individual comparisons for the diurnal house finch (from Tables 6 and 7 in Hamner and Enright 1967; and Tables 1 and 2 in Enright 1966b). B Data for three species of nocturnal rodents, recorded under identical protocol: averages for several individuals in each case. o: hamster, *Mesocricetus auratus;* Δ: deermouse, *Peromyscus leucopus;* •: deermouse, *Peromyscus maniculatus* (from Table 3 in Daan and Pittendrigh 1976b). C Intra-individual comparison for a single blinded house sparrow (derived from Fig. 4 in Menaker 1968); D Combination of inter- and intra-individual comparisons for the diurnal Siberian chipmunk (extracted from Fig. 5 in Pohl 1972); E Nine intra-individual comparisons for the pocket mouse, *Perognathus longimembris,* from an experimental protocol with 3 weeks in darkness (left column of data), 3 weeks in continuous light (140 lux $<$ I $<$ 1160 lux) and then 3 weeks thereafter in darkness (data from Lindberg and Hayden 1974). F Inter- and intra-individual comparisons for the diurnal antelope ground squirrel, *Ammospermophilus leucurus,* from Table 7 in Kramm (1971). The data shown in parts A, C, and E have not been considered in previous tabulations of species-specific responses to constant light intensity; the data in parts D and F involve cases which have been previously interpreted as showing no change in period due to light intensity

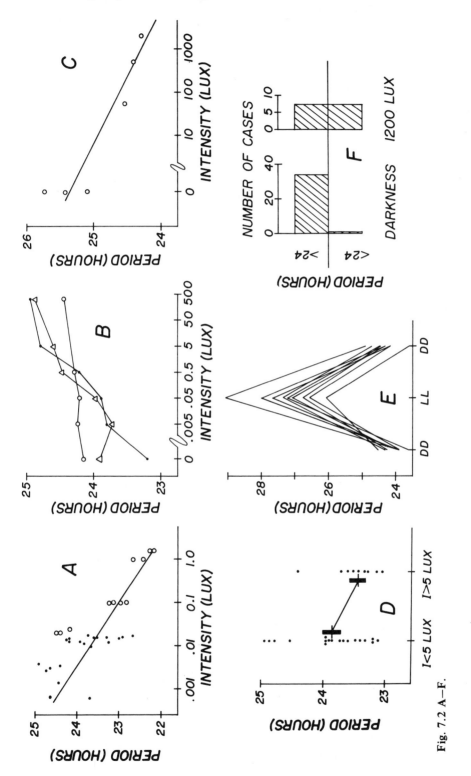

Fig. 7.2 A–F.

These qualitative differences in the manner in which constant light intensity alters circadian rhythms have been the focus of much attention in the circadian literature; interspecies comparisons of the data show a kind of consistency which was first noted by Aschoff (1958, 1961), and which has been designated Aschoff's rule (Pittendrigh 1961; Hoffmann 1965): light decreases the period of the rhythm of diurnal organisms (both birds and rodents); light increases the period of the rhythm for nocturnal species. (See Pohl 1972; Daan and Pittendrigh 1976b; and particularly Aschoff 1979, for recent summaries of the data; see App. 7.A, for a consideration of exceptions.) The ecologically correlated trends expressed in Aschoff's rule demand some sort of assumption which distinguishes between diurnal and nocturnal animals, in terms of the action of light on the pacemaker models, in order to account for the opposite ways in which light can affect free-running period under constant conditions. Since it proved convenient, for other reasons entirely (see above), to assume that the influence of light on the pacemaker is mediated by discriminator threshold, let us consider the possibility that light might decrease threshold of the discriminator for diurnal animals and increase threshold for nocturnal forms.

This assumption is not as arbitrary or ad hoc as it may at first glance seem to be. On elementary ecological considerations, one might well expect light to affect diurnal and nocturnal animals in opposite ways, being excitatory in some general sense for day-active animals and being inhibitory for those active at night. In a behavioral context, an excitatory *phasic* stimulus is usually considered to be one which evokes an immediate response; and an inhibitory phasic stimulus would be recognized by the criterion that it briefly interrupts some sort of ongoing activity. The nature of a *tonic* stimulus, however, is more easily defined in terms of induced changes in the frequency or probability of ongoing activity: excitation increases that probability and inhibition decreases it. If one assumes that the ongoing activity of interest arises due to a spectrum of continual excitatory stimuli from other sources, then an alteration in the probability of response to those stimuli can be equated, in a qualitative way, with an alteration of the threshold of the system for responses to the stimuli: supplementary tonic excitation lowers the threshold and tonic inhibition raises the threshold. In the present context, the "continual stimuli from other sources" would represent summed pacer discharge; and the interpretation is that light alters the probability that the discriminator responds to those stimuli, by lowering (diurnal animals) or raising (nocturnal animals) the discriminator threshold.

While such considerations suggest that the assumption proposed to account for qualitative interspecies differences in the action of continuous light has a certain intuitive plausibility, the practical advantages of the assumption here extend beyond observations under constant conditions. In subsequent chapters, interest will be focussed upon the effects of light-dark cycles upon the pacemaker models. At that point, we must again deal with the observation that some animals are active during the light portion of the environmental cycle and others during the dark portion; that some are diurnal and others are nocturnal. For a Coupled Stochastic System, in which light affects discriminator threshold, it would prove necessary at that point — if we had not already done so here — to assume that light lowers threshold for diurnal animals and raises threshold for nocturnal animals. No other assumption can satisfactorily account for the diurnal-nocturnal dichotomy, without fundamental restructuring of the models. In other words, we subsequently require the same assumption, proposed here to account for interspecies differences in response to constant light, in order to account for interspecies differences in responses to light-dark cycles. In Coupled Stochastic Systems, the ecological dichotomy between diurnality and nocturnality is inescapably linked (or almost so — see App. 7.A) with fulfillment of Aschoff's rule under constant conditions. A single assumption determines both sorts of behavior; one becomes a consequence of the other.

7.2 Effects of Light upon Duration and Intensity of Activity

Aschoff (1961) was also the first to draw attention to two other effects of constant light upon free-running circadian rhythms, which were summarized in what he called the circadian rule.[5] The properties of interest are the duration and the amount (or intensity) of locomotor activity within a circadian cycle: for diurnal animals, both duration and maximal intensity of activity tend to increase with increases in light intensity; for nocturnal animals, light tends to have opposite effects.

The examples of Figure 7.1 clearly illustrate this kind of result. Exceptions to these trends have been documented in nocturnal insects (data summarized in Hoffmann 1965; Daan and Pittendrigh 1976a), but observations made of birds and rodents are nearly all in good agreement with Aschoff's generalization, which, in its most general form, states that frequency of a circadian rhythm, per-cycle duration of activity and amount (or intensity) of activity all are changed in the same direction by changes in the intensity of constant light: increased for diurnal animals and decreased for nocturnal forms. Far fewer sets of data have been published on duration and amount

---

5 My terminology here conforms with that of Hoffmann (1965), who distinguished between Aschoff's rule for effects of light on period of a rhythm, and the circadian rule for other effects of light on locomotor activity; it remains anomalous, however, that one of the dozens of generalizations which might be made about circadian rhythms is considered important enough to be given the unique designation of the circadien rule.

TIME OF DAY

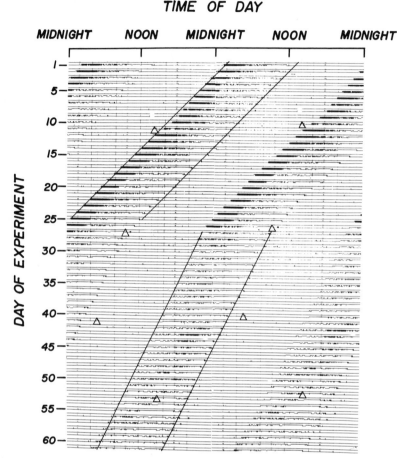

Fig. 7.3. Activity record of an isolated house finch under constant light conditions. Feeding disturbances, with no change in light intensity, indicated by *triangles*. Entire record has been duplicated, with right panel raised one day, to permit following of trends which scan more than 24 h. The sudden change in activity pattern and in period of the rhythm on day 27 may well have been evoked by the intervening feeding disturbance, but there was no change in prevailing light intensity or in any other aspect of constant conditions at that time

of activity, as affected by light, than are available about period of the rhythm, and in part this is because period is so much simpler to evaluate. The irregular pattern in ends of locomotor activity means that estimates of duration of activity are often subject to great uncertainty; and more expensive instrumentation is usually necessary for quantitative evaluation of the amount of activity.

One may well entertain reservations about whether these experimental variables can tell us anything of interest about the pacemaker itself. In contrast with the effects of light upon period of a circadian rhythm, where the pacemaker is clearly involved, it is entirely conceivable that the effects of light upon duration and amount of locomotor activity are only peripheral phenomena, having nothing directly to do with pacemaker behavior. Light might alter the animal's excitatory state, during hours of wakefulness, because of a stimulatory or inhibitory action which has no bearing on the fundamental circadian timing process. In evaluating the latter interpretation, it is useful to consider certain experiments in which an animal changes its day-to-day activity pattern suddenly and dramatically during a prolonged free-run, without any concommitant change in prevailing constant light intensity or in any other experimental variable. In a number of such cases, a simultaneous and equally striking change in the free-running period of the rhythm has been observed (Figs. 7.3, 7.4, and 7.5). Such large and sudden changes in properties of a free-running rhythm are unusual and unpredictable events, sometimes clearly associated with a single perturbation, sometimes apparently spontaneous. Their broader significance, regardless of cause, resides in the impressive *simultaneity* of changes in all three factors (period of the rhythm, duration of activity, and intensity of activity), leading to the suspicion that some single factor within the pacemaker has changed, which determines all three properties. Such cases thereby offer strong support for Aschoff's suggestion that these three facets of an activity rhythm are causally interrelated: that those changes in all three which occur in response to changes in light intensity do not represent separate, independent actions of light, some central to the pacemaker and others peripheral; but instead constitute different aspects of a single action of light upon the pacemaker itself.

The assumption that light intensity exerts its effect upon discriminator threshold leads directly to the inverse correlation, described by Aschoff, between period of a circadian rhythm and duration of locomotor activity. Exactly that result is illustrated in Figure 6.9. If constant light raises discriminator threshold for nocturnal animals, and lowers threshold for diurnal animals, then Figure 6.9 indicates that duration of wakefulness and period of the rhythm would be altered by light intensity in opposite directions, as observed, for nocturnal and diurnal animals — but this should not be surprising. The assumption that light modifies discriminator threshold, rather than entering the system in some other way, was predicated upon conformity with exactly this kind of result.

The second component of the circadian rule, dealing with the total amount and intensity of activity (note the striking changes in Figs. 7.1, 7.3, 7.4, and 7.5), is also qualitatively compatible with Coupled Stochastic Systems, if one makes the further, rather natural interpretive assumption that the intensity of locomotor activity is a function of the magnitude of *suprathreshold* input to the discriminator (cf. Fig. 5.3). Clearly, with a decrease in threshold (i.e., more light for a diurnal animal, or less light for a nocturnal

TIME OF DAY

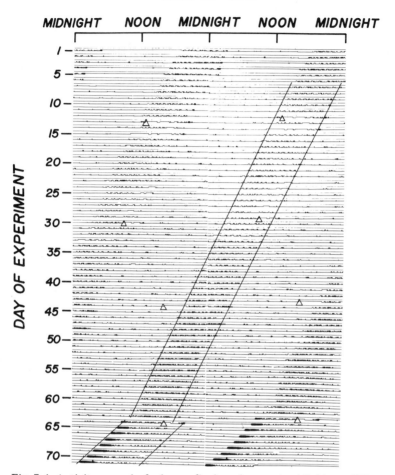

Fig. 7.4. Activity record of a house finch under constant light conditions. As in Figure 7.3, each 24-h segment of the record has been duplicated on the right half of the illustration. Feeding disturbances indicated by *triangles*. The dramatic change in activity pattern and period of the rhythm, during the last 8 days of the record, was not associated with any change in prevailing constant conditions

species), the peaks in summed pacer discharge will rise further above threshold. (Note that the absolute amplitude of the oscillation in summed pacer discharge does *not* change appreciably with threshold; the suggestion here is that the *effective* portion of pacer discharge is that amount by which the summed value exceeds threshold of the discriminator.)

It should not escape our attention, however, that the intensity of locomotor activity can differ by as much or more than an order of magnitude (e.g., Figs. 7.1, 7.3, 7.4, and 7.5), with only modest changes in period of the rhythm; and that changes in the amount by which summed pacer dis-

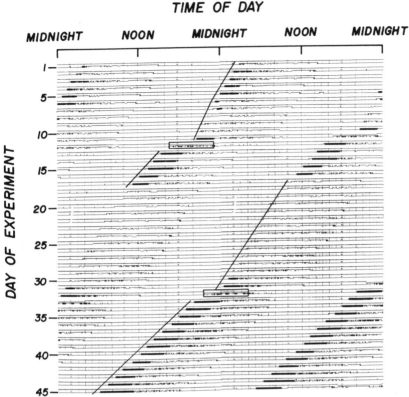

TIME OF DAY

Fig. 7.5. Activity record of a house finch under constant light conditions, except for two 6-h pulses of bright light, indicated by superimposed *rectangles*. Estimates of period of the rhythm, for portions corresponding to the series of superimposed *diagonal lines,* are: 23.38 h for days 1 to 6; 23.42 h for days 6 to 11; 23.04 h for days 12 to 17; 23.33 h for days 18 to 32; and 22.89 h for days 33 to 45. In each case the change in period of the rhythm, whether apparently spontaneous (days 6 and 17) or induced by light perturbation (days 11 and 32) was accompanied by a dramatic change in the activity pattern. In keeping with the circadian rule, the 20-min shortening of period on day 33 was accompanied by a lengthening of the duration of activity by more than an hour (duration before = 11.03 h, ± 0.17; duration after = 12.28 h, ± 0.08, where ± represents standard error)

charge exceeded threshold, in the simulations which led to Figure 6.9, involved changes which were not more than a factor of 2 to 3 throughout the entire range of period values. There is, then, a major quantitative discrepancy here. That problem could of course be remedied by postulating nonlinearity in the transformation of pacemaker behavior into amount or intensity of locomotor activity. Since such elaboration of the model, however, serves no purpose in subsequent chapters, it is perhaps best here simply to note a qualitative agreement between model and data, with quantitative disagreement which would require patchwork to remedy.

7.3 Effects of Light Intensity upon Precision

Extensive consideration was given in Chap. 6 to factors affecting precision of the pacemaker, and one aspect of those simulations becomes of interest here. Figure 6.8A indicates that the precision of pacemaker onsets is related in a nonmonotonic way to discriminator threshold; minimum variability occurs at intermediate values of threshold. Since period of the pacemaker rhythm shows a strong positive correlation with threshold (lower part of Fig. 6.9), the pacemaker will show maximum precision at intermediate values of period (Fig. 7.6B). At present, the only experimental observations with which this property of the model pacemaker might be compared consist of data obtained from a diurnal bird, the chaffinch (*Fringilla coelebs*). Those data are summarized in Fig. 7.6A, which shows the relationship between standard deviation of activity onsets and free-running period in some 60 experiments conducted at several light intensities. The scatter in the data is large, but a minimum in standard deviation (i.e., maximum precision) is nonetheless evident at intermediate values of period, as in the simulations with the model. It is encouraging to find such clear conformity between performance of the model and experimental data, for a case in which the relationship between variables is not monotonic, but instead passes through a minimum.

A second aspect of the data from these same chaffinch experiments is also noteworthy: as shown in Figure 7.6C, the ratio of variability in *ends* of activity to variability in *onsets* declined monotonically with increasing period. In this feature, as well, the type model showed behavior which corresponds well with the experimental data (Fig. 7.6D, which replots the data from Figs. 6.8 and 6.9). Again, then, there is an interesting conformity between performance of the model and experimental data.

7.4 Extremes of Constant Lighting Conditions

In the results of Figures 6.8 and 6.9, only data from simulations with threshold values between about 20% and 45% of the total number of pacers are illustrated. With the type model, a value of discriminator threshold within this range is essential to the coordinated functioning of the entire system on a long-term basis. At lower values of threshold, the pacemaker output degenerates into continuous feedback by the discriminator within a few cycles. Summed pacer discharge remains continuously above threshold, and internal coordination of the system is thereby lost. With continuous feedback, the separate pacers then oscillate independently in their accelerated state (cf. Fig. 4.5). At the other extreme, values for threshold greater than

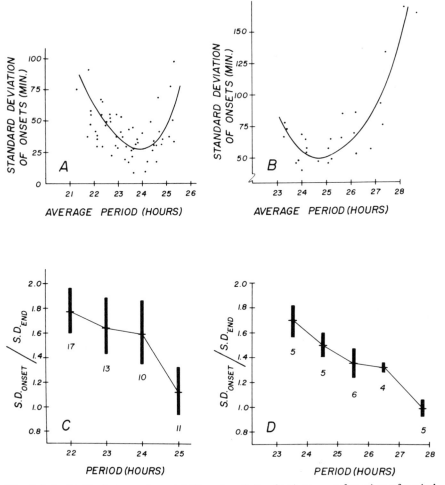

Fig. 7.6 A–D. Cycle-to-cycle variability of activity rhythms, as a function of period, compared with variability from simulations with pacemaker model. **A** Standard deviation of activity onsets in experiments at several different light intensities, as a function of free-running period, for the chaffinch (from Aschoff et al. 1971). **B** Standard deviation of onsets, as a function of period, for the type model (data from simulations of Figs. 6.8 and 6.9, replotted in format comparable with part **A**). **C** Relationship between ratio of standard deviation of ends of locomotor activity to standard deviation of onsets of activity, and free-running period, for the same experiments as in part **A**. Heavy *vertical bars* represent ± standard error of the mean, for the values of the ratio, grouped by one-hour intervals of period. **D** As in part **C**, for the computer simulations which led to part **B**

about 0.45N lead, within a few cycles, to permanent failure of the summed pacer discharge to exceed threshold; internal coordination of the ensemble is rapidly lost due to the absence of feedback, and the pacers again oscillate independently, but this time in the basic state. Thus, Coupled Stochastic Systems are stable only within a restricted range of values for threshold.

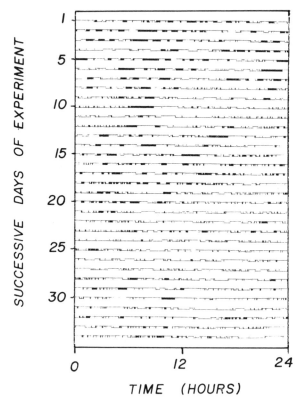

Fig. 7.7. Locomotor activity of a house finch under constant bright light (ca. 300 lux)

An experimental phenomenon, which appears to be directly comparable with the behavior of the pacemaker models under very low levels of threshold, consists in the arrhythmic behavior of diurnal birds, when placed in constant bright light. With present assumptions, bright light would correspond to extreme lowering of discriminator threshold. The light intensity necessary to induce arrhythmia in a bird varies appreciably among species, but is often on the order of a few tens of lux. All species of diurnal birds for which activity has been monitored under constant bright light show such arrythmia, which consists of essentially continuous locomotor activity, with no evidence whatever for sustained intervals of sleep (Fig. 7.7). Sensitive statistical analysis has shown that there is no detectable residual circadian variation in the amount or timing of activity, nor any indication of circadian rhythm in the birds' normally cyclic body temperature under bright lights (Binkley et al. 1972). The fact that a bird can survive many months in this state, without sustained intervals of sleep, is in itself remarkable. Beyond this, it is puzzling that persistent circadian rhythms in birds can only be obtained at constant light intensities less than about 100 lux, although normal daylight exceeds 10,000 lux; but daylight is not, of course, constant. It has previously been suggested, on an intuitive basis, that this kind of phenomenon may well represent a loss of mutual entrainment within a multi-oscillator pacemaker system (Underwood 1977; Gwinner 1978). Coupled Stochastic Systems lead to a similar interpretation: the assumption that light acts upon the pacemaker by lowering discriminator threshold for diurnal animals means that very bright continuous light ought to produce continuous discriminator feedback, which destroys mutual entrainment within the ensemble of pacers. The observed arrhythmia is thus a natural consequence, a corollary of the way the postulated pacemaker was assumed to respond to light, in order to account for Aschoff's rule in a simple way.

Unless further qualifications are invoked, the pacemaker models predict that equivalent arrhythmia should also be observed of diurnal rodents in bright continuous light — but it is not; circadian rhythmicity persists. Furthermore, if one places no additional restrictions on the relationship between light intensity and threshold, and thereby assumes that *decreases* in light might, for a nocturnal animal, also lower threshold indefinitely, then comparable arrhythmia should be observed of a nocturnal rodent in complete darkness — but it is not. In order to account for these experimental results with rodents, it must be postulated that the threshold of the discriminator "saturates" in its responses to light intensity. For diurnal rodents, we must presume that bright light is not as stimulatory as it is for diurnal birds: threshold saturates at some intermediate intensity. (Such a difference would be consistent with the fact that the effects of light on the rhythms of birds are not mediated by the eyes, although they are in mammals; see Chap. 15.) Furthermore, we must presume for nocturnal rodents that darkness does not lower threshold as far as bright light lowers it for diurnal birds: darkness is not as "stimulatory" for a nocturnal animal as bright light is for a diurnal bird. This latter interpretation would seem, in fact, to be demanded by ecological considerations alone: many nocturnal rodents retire for sleep to burrows where darkness prevails; if darkness were to stimulate such an animal to continuous activity by lowering threshold indefinitely, then the animal could not sleep in its burrow if it behaved like the model.

When threshold of the pacemaker models is placed at too high a level, mutual entrainment of the ensemble is also lost because of the absence of cyclic feedback. Such high levels of threshold might, in principle, be expected of a diurnal bird, kept in complete darkness, and of a nocturnal animal kept in extremely bright light. Phenomena which, at first glance, seem to bear out this expectation are the observations, 1) that may diurnal birds (but not the English sparrow or the European starling) are completely inactive in total darkness, and soon starve to death, because of an inability to feed; and 2) that extremes of bright continuous light often completely inhibit wheel-running activity by a nocturnal rodent. It is conceivable that both of these phenomena correspond to extreme raising of discriminator

threshold; that these conditions, if maintained sufficiently long, would lead
to progressive damping of pacemaker oscillations, and eventual loss of circadian rhythmicity. It is my intuition, however, that there may be natural
upper limits in the extent to which discriminator threshold can be forced
by extremes of light intensity or darkness: a "saturation" phenomenon resembling those described above for low-threshold conditions, which sets
limits so that the circadian pacemaker oscillation can persist undamped.
These several proposed relationships between threshold of the discriminator
and light intensity are summarized schematically in Figure 7.8.

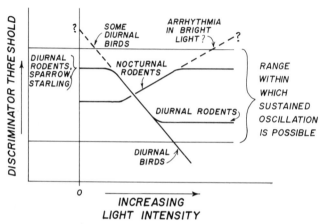

Fig. 7.8. Schematic illustration of the relationships between threshold of the discriminator and light intensity for different taxonomic groups, as proposed in text. *Solid lines*
represent relationships apparently demanded by available experimental data; *dashed
lines* represent possibilities not yet excluded by experiment

   For all experimental data and predictions to be considered in subsequent
chapters, involving light cycles and light pulses, only the middle portion of
the graph in Figure 7.8 is required: the assumption that light lowers discriminator threshold for diurnal species and raises threshold for nocturnal
species. The other portions of the figure involving saturation are required
only to account for interspecific differences observed under extremes of
constant light intensity, and for that reason, there is no need for quantitative specification of the curves shown there. The essential point of this figure is only the assumption that saturation phenomena arise at extremes of
stimulus intensity. Comparable phenomena are quite common in biological
systems, but it usually proves convenient to ignore such problems and to
concentrate on "small perturbations" as will be done in subsequent chapters here.

7.5 Ranges of Free-running Period Values

Although the free-running periods of circadian wake-sleep rhythms vary with the intensity of the constant light provided (cf. Aschoff's rule, above), the extent of this variation is relatively small. In extreme cases, light may change the free-running period by up to 4 or 5 h, as in the pocket mouse *Perognathus longimembris* (cf. Fig. 7.2); and a similar value (ca. 3.5 h) has been obtained with the chaffinch (Aschoff et al. 1962); but usually the intraspecific range of free-running periods inducible by constant light is on the order of only 1 to 2 h. Computer simulations have demonstrated that Coupled Stochastic Systems are similarly restricted to a relatively narrow range of period values, when threshold of the discriminator is assumed to respond to changes in light intensity.

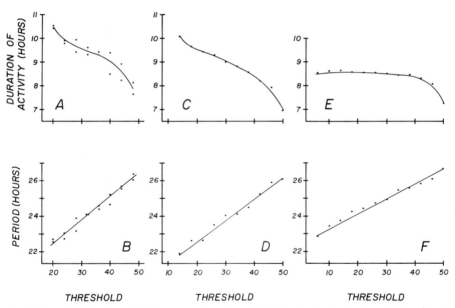

Fig. 7.9 A–F. Relationships between duration of feedback (parts **A**, **C**, and **E**) or period of the pacemaker rhythm (parts **B**, **D**, and **F**), and threshold of the discriminator, within the range of period values for which sustained oscillation is possible. For parts **A** and **B**, beta was assigned a value of 1 h, and all other parameters were as in the type model; for parts **C** and **D**, beta was again assigned a value of 1 h, and gamma was assigned a value of 1/24; for parts **E** and **F**, beta was assigned a value of 1 h, gamma a value of 1/24, and alpha a value of 1/4 h

Changes in threshold in the type model can only alter the period of the pacemaker system by 4 to 5 h (Fig. 6.9); more extreme values of period are not realizable because coordinated behavior of the system deteriorates at

more extreme values of threshold. A similar restriction apparently applies
to all other models defined by the equations of Chap. 4. The issue has not
been examined exhaustively, but many other specific examples of the class
have been investigated by varying parameters from their type-model values,
and in none of them has a larger range of periods been obtained due to
changes in threshold than that found for the type model. Neither α, β, γ,
nor N apparently has any appreciable influence on the range of achievable
period values (cf. Fig. 7.9); and ϵ can be varied from 3 to 12 h without increasing that range (Fig. 7.10). Values for $\bar{\delta}$ appreciably greater than 0.5,

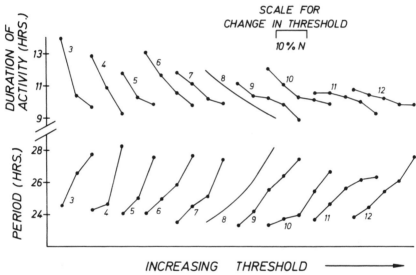

Fig. 7.10. Influence of threshold upon duration of feedback activity (*upper part*) and
period of the rhythm (*lower part*), for various values of ϵ. Threshold was varied in steps
of N/20, within the range at which sustained oscillations persisted. *Small numbers* next
to the curves (3 to 12) indicate the value of ϵ in hours. Curves for ϵ = 8 h taken from
Figure 6.9. Simulations with 100 pacers, with all other parameters as in the type model

however, or values of ϵ greater than 12 h can greatly *reduce* the range of
realizable period values (Figs. 7.11 and 7.12). Furthermore, the propositions
of Figure 7.8, which involve saturation (an upper or lower limit on the extent to which threshold can be altered by light) also imply potential reduction in the range of period values produced by changes in light intensity.
Hence, Coupled Stochastic Systems not only account for the empirical observation that constant light can alter free-running period by not more than
4 to 5 h, but also are compatible with the observation that many species
show an even smaller range. This correspondence between model and experimental data is of significance primarily in that it demonstrates an additional advantage of assuming that light affects the discriminator threshold

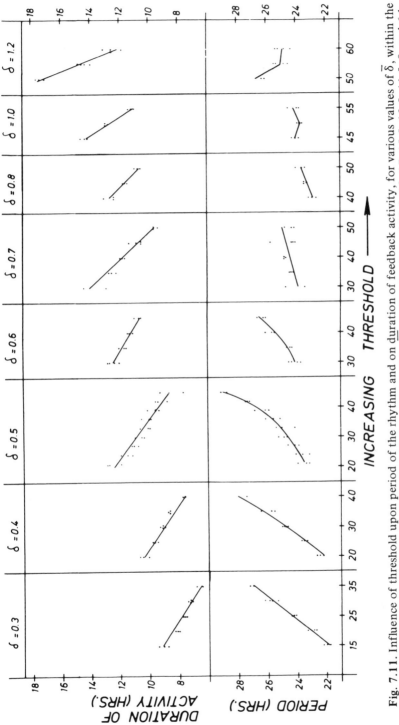

Fig. 7.11. Influence of threshold upon period of the rhythm and on duration of feedback activity, for various values of δ, within the range of threshold values at which sustained oscillation was possible. $\overline{X}_b$ assigned values of 19.5, 18, 17, 15.5, 13.5, 11.5, 9, and 6 h, for δ values of 0.3, 0.4 ... 1.2, respectively. All other parameters as in the type model. Values for $\delta = 0.5$ taken from Figure 6.9. Note that with a value of 1.2 assigned to δ, the period consistently decreased with increase in threshold

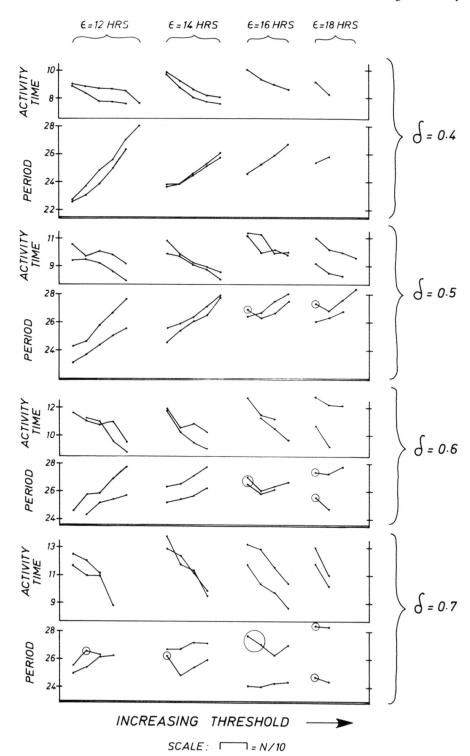

Fig. 7.12.

of the models. If light were presumed instead to act upon some other parameter of the system, e.g., to alter $\overline{X_b}$, a limitation on free-running periods of the system would have to be artificially imposed by an additional ad hoc assumption.

7.6 Summary

The central conclusion to be drawn from this chapter is that several different aspects of the experimental data obtained under constant conditions can all be simply accounted for by making the assumption that light monotonically alters the threshold of the discriminator. In view of the fact that qualitatively different responses are obtained from day-active animals and from those which are night-active, the further specification is required that light lowers threshold for diurnal animals and raises threshold for nocturnal animals. With this postulated mode for the action of light, Coupled Stochastic Systems can satisfactorily reproduce the following phenomena, which have been observed in experiments in which the intensity of constant light was varied: (1) those changes in period of circadian rhythms which are summarized in Aschoff's rule, observed in essentially all nocturnal and diurnal vertebrates; (2) those commonly observed changes in per-cycle duration of wakefulness, which are summarized in the circadian rule; (3) the greater precision in onsets of locomotor activity than in ends of activity, which is typical of most diurnal and nocturnal species, as well as the decrease in the ratio between variability in ends and variability in onsets, with decreasing light intensity, documented in the chaffinch; (4) the existence of a maximum in precision of activity onset at intermediate values of period of the circadian rhythm, documented in the chaffinch; and (5) the observation that free-running period of a circadian rhythm cannot be altered by more than 4 to 5 h by changes in constant light intensity.

Fig. 7.12. Influences of threshold upon period of the pacemaker rhythm, and on duration of feedback activity, for various combinations of ϵ and $\overline{\delta}$, the Excitation and the Discharge Coefficient parameters. $\overline{X_b}$ assigned as in the calculations of Figure 7.11. α, β and γ had type-model values. Threshold was varied in steps of N/20. *Lines* connect series of calculations in which identical parameters ($z_{i,\beta}$ and $z_{i,\gamma}$) were assigned to the set of pacers, and only threshold was varied; the two sets of data, i.e., two lines in a given block, were determined from simulations with different pacemaker ensembles, arising from different arrays of values for $z_{i,\beta}$ and $z_{i,\gamma}$, drawn from the Gaussian distribution. Instances enclosed in *open circles* represent cases in which Aschoff's rule was violated, in that period of the rhythm became longer at lower values of threshold. Some of these, which were not duplicated in calculations with a second set of pacers (e.g., $\epsilon = 16$, and $\epsilon = 18$ h, at $\overline{\delta} = 0.5$), may represent chance events, but other cases clearly involve reproducible phenomena

In addition, the models conform qualitatively to the widely observed effects of light intensity upon per-cycle amount of locomotor activity under different levels of constant light intensity. Unless further assumptions are invoked, however, the quantitative agreement is poor in this latter case. The supplementary and more specific assumptions shown in Figure 7.8 permit the models to reproduce satisfactorily the consistently observed arrhythmia of diurnal birds in bright light, and the lack of this phenomenon in mammals. As we shall see in subsequent chapters, the assumption that light affects the threshold of the discriminator in Coupled Stochastic Systems proves also to be a remarkably economical way of accounting for a variety of experimental results in which an animal has been subjected to light cycles and light pulses.

Appendix 7.A. Aschoff's Rule: Models with Other Parameter Values

The data presented in Figures 7.1 and 7.2 are all cases of the sort which provide the foundation for Aschoff's rule: the period of the rhythm is shorter for diurnal animals maintained in brighter light, and longer for nocturnal animals. This notion, which was initially formulated on the basis of very few examples, has subsequently turned out to be a surprisingly good empirical generalization; nevertheless, occasional observations to the contrary have been reported, and such exceptions give rise to this appendix.

The first *kind* of exception to the rule has been reported for the owl *Tyto alba* (Erkert 1969). On the assumption that an owl should be classified as nocturnal, one would expect, according to Aschoff's rule, that the period of the owl's circadian rhythm would be longer in bright light than in dim light; and that result was indeed obtained for light intensities greater than about 0.5 lux. At light intensities less than 0.5 lux, however, the period was also longer than that minimum value obtained near 0.5 lux. Correlated with this result, both duration of activity and intensity of activity were maximal at about 0.5 lux, decreasing at both brighter and dimmer intensities. This kind of exception to Aschoff's rule can be readily reconciled with the models: since frequency of the rhythm, duration of activity, and amount of activity remained consistently positively correlated, a straightforward interpretation for the owl data is that threshold of the discriminator may be nonmonotonically related to light intensity, with minimum threshold at about 0.5 lux. That interpretation implies that both exceedingly dim light and exceedingly bright light are "inhibitory" for the owl — and this seems sensible to me, since no matter how efficient its visual system, an owl would presumably hesitate to fly or to show intense activity in total darkness. A prediction which follows from the present models thus modified, as a test of this interpretation, is that in a light-dark cycle, in which the "daytime" intensity is on the order of 0.5 lux, the owls should become diurnal, in the sense that their wakefulness would coincide with the brighter portion of the cycle.

Data from experiments with the rhesus monkey (Martinez 1972) constitute a very similar kind of "violation" of Aschoff's rule: minimum values of period were obtained at between about 2 and 10 lux, with greater free-running periods measured at both brighter and dimmer intensities.[6] Other data from these same experiments (ibid., Fig. 4B) suggest that the light intensity, at which maximal per-cycle activity occurred, roughly coincided with that at which minimum period of the rhythm was observed. As with the data for the owl, these results suggest that both very bright light and very dim light may be inhibitory for the rhesus monkey — although, of course, the "optimum" light intensity for the monkey is much greater than for the owl. A similar interpretation, invoking a nonmonotonic relationship between light intensity and threshold, would account for such data. The suggestion that very bright light might be somewhat inhibitory for a normally diurnal species is not, in my opinion, inconsistent with the dense-jungle habitat of rhesus monkeys. As in the case of the owl, an obvious experimental test of the proposed interpretation suggests itself: in a 24-h light cycle, in which the daytime intensity is, say, 1000 lux and the night-time intensity is, say 5 lux, the monkeys should be expected to behave as "nocturnal" organisms, sleeping during the bright-light phase of the cycle. Furthermore, subsequent free-run experiments are predicted to show that this inverted phase relationship reflects pacemaker behavior, and not simply exogenous suppression of activity by bright light.

Other reported violations of Aschoff's rule in higher vertebrates consist of two cases reported for diurnal mammals, one in which observed free-running period of the activity rhythm was apparently little affected, if at all, by constant light (the tree shrew, *Tupaia belangeri*) and another in which period increased with increasing light (the palm squirrel, *Funambulus palmarum:* Pohl 1972). The case of the *Tupaia* deserves special note; although no appreciable effect of light upon free-running period was noted, other remarkable effects of light intensity upon the activity rhythm were observed. Because of those effects, the case is important enough that it will be considered subsequently in some detail (Chap. 14).

The data from the palm squirrel (Pohl 1972) are an extreme anomaly among birds and rodents, a puzzling case which is therefore of particular

---

6 It appears that the average period value plotted for zero intensity, in Figure 3 of Martinez (1972) has been miscalculated by about 10 min, based on the values given in Table 2 of that paper. With this correction, the intermediate minimum in period value at intermediate intensities becomes much more evident.

interest here. This Asiatic rodent is diurnal (light-active) both in the laboratory, and in its natural environment (ibid.). Nevertheless, the overall data reported on its behavior under constant conditions are quite consistent in indicating that period of its circadian rhythm usually increased in brighter light, over the entire range from 0.5 lux to 1000 lux. There is no clear basis in the data for proposing a minimum value of free-running period at intermediate intensities, as for the owl and the rhesus monkey, and furthermore, the amount of activity per cycle, shown by the palm squirrels in these experiments, was *positively* correlated with free-running period, also increasing in brighter light. Light seems to have stimulated locomotor activity, and yet to have increased period of the rhythm. Such a result gives rise to the question whether the models must be fundamentally revised in order to account for such anomalous experimental results.

That question is, in fact, of considerable theoretical interest, independent of the data from the palm squirrel: if conformity with Aschoff's rule is only a property of the type model, if other choices of parameter values can readily lead to different behavior, then that portion of Chap. 7, which placed such emphasis on the fact that the models can provide a ready explanation for this phenomenon, would have been seriously incomplete. This issue has been explored in supplementary simulations, which indicate that while most choices of parameter values lead automatically to the trends described in Aschoff's rule, certain sets of extreme values can be found which lead to exceptional behavior by the pacemaker model.

The values assigned to α, β, γ apparently have no bearing on Aschoff's rule; in all cases examined, in which only these parameters were altered, the period of the pacemaker was positively correlated with the level of threshold (examples given in Fig. 7.9, parts B, D and F). It should not escape our attention in Figure 7.9E, however, that when α was assigned a small value, in conjunction with values for β and γ which were appreciably smaller than in the type model, there was, over a wide range of threshold values, essentially no change in the duration of feedback activity with change in threshold. This would correspond to a situation in which changes in light intensity affect the period of the wake-sleep rhythm (Aschoff's rule) without altering the duration of wakefulness (violating the circadian rule). Hence, interspecific differences in the magnitude by which duration of wakefulness changes with unit change in period could reflect difference in the magnitude of α, the stochastic, cycle-to-cycle variability of the pacers.

As shown in Figure 7.10, the magnitude of feedback, ϵ, can be varied from 3 to 12 h without any reversal in the influence of threshold on the period of the pacemaker rhythm, or on duration of wakefulness; Aschoff's rule and the circadian rule are consistently fulfilled. As we shall see in subsequent chapters, other kinds of data apparently are best accounted for by the assumption that ϵ has a relatively large value (ca. 12 h) for nocturnal rodents like the hamster and the flying squirrel, and a relatively small value (ca. 5 h) for diurnal birds like the sparrow and the house finch. In view of Figure 7.10, one might therefore expect that the duration of activity would be altered less, in rodents than in birds, by changes in threshold (and light intensity) which induce a given alteration in free-running period. This expectation is in satisfactory agreement with experimental observations (cf. Fig. 7.1).

In order to induce pacemaker behavior in which Aschoff's rule is not fulfilled, the parameter $\bar{\delta}$ proves to be critical. As shown in Figure 7.11, the larger $\bar{\delta}$ is, the greater the effect of threshold on the duration of activity, and the smaller the effect on period of the system. When $\bar{\delta}$ was assigned a value as large as 1.0 or 1.2, the period of the system decreased slightly with an increase in threshold, violating Aschoff's rule. Nevertheless, at these values of $\bar{\delta}$, the duration of activity still decreased strikingly as threshold increased, in conformity with the circadian rule.

Other simulations, in which $\bar{\delta}$ was varied, in combination with larger values of ϵ, show that Aschoff's rule can also be violated by the pacemaker models, even when $\bar{\delta}$ is less than 1.0, provided that ϵ is assigned a very large value. Several combinations of these two parameters have been found (Fig. 7.12) with which there was either no consistent effect of threshold on period, or very modest decrease in period with increasing theshold.

An initial overview of the behavior of the models in Chap. 6 led to the expectation that period of the system should consistently increase with increasing threshold, based on the intuitive argument that attainment of a higher threshold by summed pacer discharge should require a longer waiting time. A complication arises, however, whenever the sum of the duration of feedback activity and the magnitude of feedback (ϵ) approaches or exceeds the expected length of the total cycle. In this circumstance, the accelerating influence of ϵ, during the latter portion of the activity time, extends through the entire inactive phase of the cycle, and therefore affects many of those pacers which would ordinarily be involved in the subsequent onset of feedback. The expected effect of threshold on period can then be reversed.

Summing up briefly the considerations in this appendix, most combinations of parameters which I have investigated for the models lead naturally to fulfillment of Aschoff's rule; and most experimental data describing the effects of light on the wake-sleep cycle of birds and rodents show similar trends. Certain combinations of extreme parameter values, however, permit the models to show opposite behavior; and occasional instances of experimental data showing opposite behavior have also been reported. Some of the anomalous experimental data (the owl and the rhesus monkey) can be accounted for by the hypothesis that discriminator threshold in such species has a nonmonotonic relationship with light intensity. Tests of that interpretation are proposed above. However, in at least one other case (the palm squirrel), another explanation is required. If the models are to account for these latter data without fundamental revision, one must suppose that the circadian pacemaker of the palm squirrel is characterized by large values of both $\bar{\delta}$ and ϵ. That interpretation can also be tested. As described subsequently (Chap. 13), such extreme parameter values lead to novel predictions for the expected behavior of the palm squirrel, when subjected to single pulses of bright light.

Chapter 8
A Brief Detour: Further Thoughts About the Discriminator of the Models

8.1 The Magnitude of Feedback

The behavior of the pacemaker, as considered up to this point, involves an acute nonlinearity in the properties of the discriminator: subthreshold input by the pacer ensemble is assumed to evoke no output whatever, and suprathreshold input to lead to full, maximally accelerating feedback: a simple switching operation. This mode of action permitted formulation of coupling in the model in Chap. 4 with only two parameters, θ and ϵ. One can find instances in the vertebrate nervous system of somewhat comparable all-or-none signal tranformation, commonly known as gating of responses, but in the majority of interactions of nerve cells, there is a continuous, quantitative relationship between input and output, between stimulus intensity and magnitude of the response. The striking nonlinearity incorporated into the pacemaker may thus appear to be extremely unrealistic. What would be the consequences of making feedback by the discriminator a more continuous function of pacer input? This question is examined here, as a separate issue from those in Chap. 6, because it has ramifications which extend beyond the class of models of primary interest; and because certain matters considered in Chap. 7 will assist evaluation of the possibilities.

For the formulation of a continuous relationship between input to the discriminator (summed value from all discharging pacers) and feedback output, a wide variety of alteratives can be envisioned. The simplest course would be to assume linearity (cases I and III in Fig. 8.1A): a direct proportionality between summed pacer discharge and amount by which recovery interval of the pacers is shortened. The proportionality indicated by line I in Figure 8.1 proved to be entirely unsatisfactory; with this kind of feedback as a modification of the type model, the system behavior consistently deteriorated into an essentially random distribution of pacer phases within a few cycles. The alternative of line II, in Figure 8.1A, truncated at feedback of 8 h, proved to be only slightly better; some simulations led to completely chaotic system behavior, and those which did not were typically very imprecise in the times at which summed pacer discharge exceeded threshold. (With N = 100, and threshold for activity onset of 30 pacers, standard deviations of 54, 114, 141, 162, and 167 min were obtained in

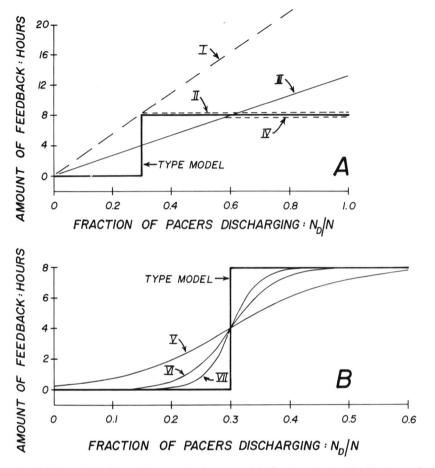

Fig. 8.1 A and B. Alternatives to the type model, for the relationship between input and output of the discriminator. **A** linear and truncated linear formulations; **B** sigmoid-curve formulations, all based on the "logistic" equation, with a saturation value of 8 h, and half-saturation at 0.3N

the "successful" simulations, compared with an average of 50 min for the type model.) Linear alternative III in Figure 8.1A, with a much shallower slope, led to appreciably better system performances; standard deviation of threshold crossings averaged about 68 min (Fig. 8.2A). Truncating this line so as to limit feedback to a maximum of 8 h (alternative IV in Fig. 8.1A) did not appreciably alter that outcome (Fig. 8.2A).

At first glance, the alternative indicated by line III might seem to be particularly attractive, since the equation for the amount of feedback requires only one parameter, the slope of the line in Figure 8.1A, rather than the two parameters, ϵ and θ, required in the type model. That interpretation, however, overlooks the fact that any such linear alternative requires

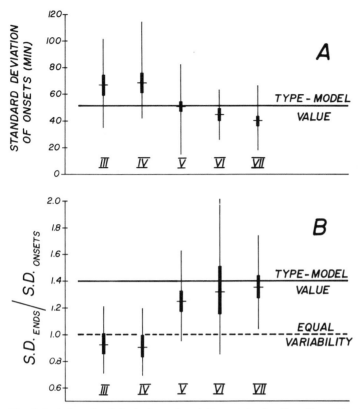

Fig. 8.2. A Standard deviation of activity onsets, for the alternative formulations of the relationship between input and output of the discriminator, as defined in Figure 8.1. The *Roman numerals* beneath the data summaries correspond to the cases illustrated in Figure 8.1. The *horizontal lines* represent the mean values from 10, 9, 20, 10, and 20 simulations, for cases III, IV, V, VI and VII, respectively; the *wide vertical bars* represent ± one standard error of the mean; the *narrow vertical lines* represent the range of observations. **B** The ratios of variability in the ends of activity to variability in onsets of activity, for the simulations in part **A**. *Horizontal lines* represent the means of the simulations (calculated on log-transformed data); the *wide horizontal bars* represent ± one standard error of the mean; the *vertical lines* represent ± one standard deviation of the observations

a second parameter to link pacemaker behavior with experimental data on locomotor activity, that is, to specify which portion of a discriminator cycle corresponds to wakefulness and which to sleep (see Chap. 7). One way or another, some parameter which is conceptually the equivalent to a threshold is, in my opinion, inescapable in linking the continuous behavior of the pacemaker, as expressed in summed pacer discharge, with records as discontinuous as those of Figures 2.1 to 2.4; none of the alternatives of Figure 8.1A is therefore more economical with parameters than the type model.

In addition to being no better than the type model in that respect, the linear-feedback alternative of line III (or line IV) is somewhat poorer in terms of the resulting precision of system performance. With a standard deviation of onsets nearly 40% greater than the type model, either of those alternatives would require nearly double the number of pacers necessary in the type model to achieve a given level of precision. The suggestion, on this basis, that a linear-feedback alternative would be "inefficient" involves, however, an issue which cannot be compared with experimental data since the number of pacers in the ensemble cannot be evaluated. A more telling argument against this kind of linkage involves the relationship between variability in onsets and in ends of simulated locomotor activity. If one assumes, as we have throughout, that threshold for the discriminator, in units of number of pacers discharging, remains constant (i.e., that the discriminator is a truly passive element), then onsets of activity for the linear alternative of line III (or line IV) are essentially equivalent, in their variability, to ends of activity (Fig. 8.2B). This aspect of system behavior would conflict strikingly with the experimental data on activity rhythms: as discussed in Chap. 7, one of the most typical features in locomotor-activity recordings from birds and rodents is that onsets of activity are appreciably more predictable in their timing than are ends. Thus, the type model, in addition to having greater precision than a linear alternative, can account for a consistent aspect of the experimental data with which a linear-feedback alternative is apparently not compatible, unless additional assumptions are invoked.

Figure 8.1B illustrates several quantitatively different curvilinear alternatives for the relationship between input and output of the discriminator, all of them sigmoid curves which saturate at 8 h feedback, the type-model value for feedback. As shown in Figure 8.2A, formulations of this sort can lead to precision in system output which is as great as (line V) or even appreciably greater than (line VII) that of the type model. Since many kinds of nervous elements, both primary receptors and connective neurons, show responses to input which have qualitative resemblances with these nonlinear formulations, the models might well be made more realistic by incorporation of such a relationship. The subsidiary argument of efficiency (fewer pacers for equivalent precision) can also be invoked, as favoring a formulation such as that of line VII. Note further (Fig. 8.2B), that the relative variability of simulated onsets and ends of activity, in all three curvilinear formulations considered, is roughly equivalent to that of the type model.

The specification of such an S-shaped curve, however, requires three parameters, rather than the two of the type model (ϵ and θ); and a fourth parameter would be needed for the threshold value which relates discriminator cycles to locomotor activity. (Threshold was arbitrarily set at 0.3N in order to caclulate estimates of variability shown in Fig. 8.2.) Since as far as I have been able to determine, a two-parameter formulation for feedback,

such as that of the type model, permits the postulated pacemaker to describe the experimental data adequately, it seems appropriate to me at this
stage merely to note the consequences of more elaborate formulations,
without incorporating them into the models.

8.2 Stochastic Variation in Amount of Feedback

The various curves in Figure 8.1 describing feedback as a function of
discharging pacers, as well as the simulations leading to Figure 8.2, all involve a proposed *deterministic* relationship between input and output of
the discriminator. In Chap. 6 the consequences of temporal unreliability in
the input-output operation of the discriminator of the type model were considered in terms of stochastic variation in threshold of the discriminator.
Here we should briefly examine the related issue of temporal unreliability
of the discriminator, when its function is not an on-off response but depends upon a continuous relationship like those of Figure 8.1B.
    Let us consider, for the moment, curve VII of that illustration; and as-
sume that moment-to-moment fluctuations in the input-output relationship of the discriminator are so large that feedback, for a given number of
discharging pacers, can vary between ± 25% of its expected average value.
When N_D = 0.3N, for example, intensity of feedback can thus range between 3 and 5 h. Note that for curve VII, 3 h feedback corresponds to the
expected average output when N_D is approximately 0.29N; 5 h corresponds
approximately to 0.31N. As shown in Chap. 6, cycle-to-cycle variations of
N_D with a standard deviation as small as ± 0.01N are to be expected for
the type-model pacemaker only when the ensemble is larger than about
1700 pacers. In the case of curve VII, therefore, the stochastic inter-cycle
variation in N_D would dominate in determining precision of pacemaker
output for ensembles smaller than about 1700 pacers, even when the input-
output relationship of the discriminator is assigned a stochastic variation
of ± 25%. For a curve steeper than that illustrated in case VII, even larger
random fluctuations in the input-output relationship would be tolerable
without detectably affecting precision of the pacemaker ensemble, unless
N becomes very large.

8.3 Can the Discriminator be Eliminated?

The fact that a continuous relationship between input and output of the
discriminator is not only a conceivable property of a Coupled Stochastic
System, but also leads to apparently satisfactory performance in other re-

spects, calls attention to an important question: Is the discriminator a necessary element in the pacemaker models? If the discriminator is not needed as an on-off switching element, but serves only to integrate pacer input and transform that input in a somewhat nonlinear manner, there is in fact no logical demand that these operations be performed by a discrete structural element.

The connective scheme between pacers and discriminator illustrated in Figure 4.3 (p. 35) is an elementary kind of "wiring diagram" by means of which pacer discharge might be integrated and transformed; but synaptic interconnection of this sort − while probably the more common mode of interaction in the vertebrate nervous system − is by no means the only possibility. One might well entertain the suggestion, for example, that the pacers of the entire ensemble are all linked in their behavior by means of a neurohormone, secreted by each pacer during its discharge phase, a neurohormone which serves to excite all nondischarging elements in the ensemble. In this kind of situation the discriminator could be eliminated as a structural component; its role could be entirely fulfilled by the magnitude of the general excitatory pool, which would then serve to alter recovery interval of each pacer.

Note, however, that the effect of such an excitatory pool on the period of each pacer should be distinctly nonlinear. As indicated above, certain features of the experimental data demand that this relationship, if not a step function as assumed in the type model, at least be a somewhat sigmoid function such as those shown in Figure 8.1B. In addition, some sort of elaboration would be required in order to formulate the manner in which light affects the system; without a discriminator and its threshold, light would have to act more directly on each pacer, in a manner which alters period of the pacemaker and other properties of its output as well. No attempt will be made here to evaluate how such operations might be accomplished by plausible physiological substrate, in the absence of a discriminator. (See, however, Chap. 15 for a related interpretation in the context of the pineal organ of birds.) These possibilities have been raised primarily to indicate certain limitations which have been rather arbitrarily imposed on Coupled Stochastic Systems.

8.4 Threshold and Feedback: an Extreme Case

Given an ensemble of pacers capable of independent oscillation, which differ appreciably from each other in intrinsic period, an extremely simple form of mutual entrainment might be attained if discharge from even a single pacer were sufficient to trigger each other element in the ensemble to discharge immediately thereafter. In terms of the kinds of models considered previously, this would be the equivalent of setting θ at a value of one pacer discharge, and assigning a very large value to ϵ, the Excitation produced by feedback.[7]

In that kind of system, the fastest of the pacers in the ensemble would assume the role of leader, and all others would be reduced to follower status. In order to accomplish such mutual entrainment, it would not be required that discharge by a pacer have a duration which is an appreciable fraction of its cycle — as is necessary when threshold for feedback depends upon simultaneous discharge by some significant fraction of the ensemble. Hence, this extreme kind of system could be formulated without the parameters $\bar{\delta}$ and γ; nor would a discriminator be a conceptually useful structural element. This is, then, a simpler kind of model than those of interest here, and constitutes a system comparable, in principle, with that underlying the endogenous rhythm of vertebrate heartbeat. There, indeed, the fastest pacer in the system leads all others, by triggering their immediate response.

As indicated in Chap. 3, however, the rhythm in heartbeat is by no means a particularly precise one; in spite of mutual entrainment of an ensemble of independent oscillators, the observed precision is no greater than that sometimes found in single spontaneous neurons. Brief consideration indicates why this might be expected: the precision of the fastest, leading pacer will be a dominant determinant of system precision. Rare and extreme events affecting this single component in the system will affect cycle length of the entire ensemble, whereas in the models defined in Chap. 4, such rare events are averaged out in the attainment of discriminator threshold. Detailed consideration (App. 8.A) of a system with a threshold of one pacer, in which all pacers are potentially equivalent in their capacity to play the role of "leader," indicates that only very modest improvement in system precision is attained by this kind of triggering system, even with a huge ensemble of elements.

Mutual entrainment produced by a single leader or "dictator" serves, in one sense, to confer reliability to the pacemaker for heartbeat: if the leading pacer should become defective, or fail to fire in even a single cycle, then a replacement is immediately available; but reliability in the sense of

---

7 In Chap. 7 simulations with many combinations of parameters consistently indicated that Coupled Stochastic Systems become unstable when threshold is sufficiently reduced. The case considered here is a conspicuous exception to that generalization; it involves the special situation in which unentrained pacers are completely absent. It is those unentrained pacers which contribute to chaotic system performance at low values of threshold in the ordinary case. They can be eliminated only if threshold is set at precisely one pacer, and ϵ is made very large.

temporal precision is not readily achievable by that mode of interaction. A further disadvantage, for present purposes, arises in trying to compare the behavior of such a pacemaker with the circadian data of interest. For example, I have been unable to devise any simple way in which light might affect a pacemaker system of that sort, which would permit it to account for the experimental data on period and duration of wakefulness considered in the preceding chapter; nor is there any obvious way by which such a system could account for experimental data on synchronization, to be considered in the next chapter. Hence, this extreme case will not be further evaluated as a candidate model for the wake-sleep pacemaker; it has been described here primarily for the sake of completeness, and I consider it to be a degenerate instance of Coupled Stochastic Systems.

8.5 Inhibition vs Excitation

Cogent reasons were given in Chap. 4 for the assumption of the models that discriminator activity should have only monotonic effects upon individual pacers of the ensemble; but the decision was made without any proposed justification to deal only with excitation, rather than inhibition. It is worth noting at this point, however, that one *can* reformulate the postulated physiological interactions which underlie performance of the pacemaker models entirely in terms of inhibition, without changing any aspect of prediction. This surprising situation arises because there is no mathematical difference in the models between increase in excitation, and decrease in inhibition.

To see how this is so, let us suppose that the discriminator exerts an inhibitory effect upon each pacer in the ensemble: it lengthens the recovery interval by an amount expressible by the parameter ϵ. Assume further that the output of each pacer contributes to an input sum for the discriminator, which can potentially inhibit its activity: as soon as that sum exceeds threshold θ, the discriminator is sufficiently inhibited to become silent. At that time, all nondischarging pacers would be released from inhibition of the discriminator: the equivalent of an acceleration which shortens the recovery process by ϵ hours. If wakefulness and locomotor activity are then presumed to ensue whenever the discriminator is silent, whenever it has been sufficiently inhibited by pacer discharge, then all properties of the system performance which might be directly compared with existing experimental data remain preserved in complete detail. Note that the coupling in such a system would, in the last analysis, remain excitatory, in the sense that discharge by one pacer increases the likelihood that another would begin to discharge.

To maintain equivalence, one must not tamper with the previous interpretation, that for a light-active animal, light lowers the discriminator threshold θ, and for a nocturnal animal, light raises discriminator threshold. Here,
there is a modest conceptual complication; light would make an active,
positive contribution toward inhibition of the discriminator in a light-active
animal, and have opposite action on a dark-active animal. It may prove
semantically confusing to demand that light is inhibitory for a diurnal animal and stimulatory for a nocturnal one, but this is a minor matter in terms
of plausible physiological substrate.

While all details of performance of the model, in terms of wakefulness
and sleep, activity and inactivity, are preserved in this conceptual transformation of the model, there are several distinctions in terms of what might
be directly measurable in the animal's central nervous system. The most important of these is that the discriminator would be active − presumably in
the sense of a continuous train of nerve impulses − only during the hours
when the animal is *inactive.* Summed pacer discharge would have high values during hours of wakefulness, and low values during sleep; and the discriminator would assume the role of an inhibitory center associated with
the sleeping state: in some senses, an active generator of sleep.

All descriptions of behavior of the models in subsequent chapters ignore
this conceptual transformation of the basic class of models under consideration; subsequent search by neurophysiologists, however, for the morphological counterparts of the components of the pacemaker postulated here may
eventually indicate that this transformed description of the system com-
ports better with real physiological substrate. (See Chap. 15.)

Appendix 8.A. Pacemaker Ensembles with a Threshold of One: A Revolving Dictatorship

Given a multi-oscillator ensemble in which the first element which reaches
a critical stage immediately triggers all others, it is self-evident that if one
element is intrinsically appreciably faster than all others, its cycle-to-cycle
precision will determine the precision of the entire ensemble. If many elements of the ensemble, however, are quite similar in intrinsic recovery rate,
so that the role of leader were to be frequently exchanged, the expected
consequences for precision of the ensemble are less transparent.

Consider, therefore, the following as an extreme case: an array of one
million pacers, each with identical recovery rate (cf. Fig. 4.1), each beginning its recovery process at an identical time and each subject to an identical noise process which can be represented by a one-per-unit-time Gaussian
variate with a standard deviation of σ. Interest then focusses upon the prob-

ability that *any one* of the pacers will be triggered by the noise process in a given unit of time. Because the component pacers are identical, and are all reset to zero at the same moment, the ensemble behavior will be equivalent to that of a single pacer of this same sort, which is subjected to one million independent "events" per unit time, from a Gaussian distribution with standard deviation σ. The largest of these events is the one which will determine whether any pacer has reached a critical state, so that it can then trigger the whole ensemble.

Given a Gaussian distribution with a standard deviation of σ, we therefore ask, "What is the distribution of largest single variates to be expected in an array of independent samples of size one million? " The median of these extreme values can be calculated as follows: $(1 - x^{1,000,000}) = 0.5$, where x is the probability that an event as large or larger than this median would *not* occur in any single trial. Solving for x gives a probability of 0.999999307, corresponding to a Gaussian variate of 4.83 σ. Similarly, the 95% confidence limits for these extreme values can be determined by setting $(1 - x^{1,000,000})$ equal to 0.975 and 0.025, leading to Gaussian variates of 4.48 σ and 5.45 σ, respectively, for the single events. Because the median (4.83 σ) is appreciably less than the calculated midrange [(4.48 + 5.45)/2], it is evident that these extreme values would have a somewhat asymmetrical distribution. Nevertheless, as a first approximation, one could treat the distribution of extreme values as having a mean of about 5 σ and a standard deviation of about 0.242 σ [i.e., (5.45 − 4.48)/4, a calculation chosen since the 95% confidence limits of a Gaussian distribution represent about ± two standard deviations]. In fact, because of the asymmetry, the effective standard deviation would be somewhat larger than 0.242 σ.

The mean value of this skewed distribution of extreme values from samples of a million Gaussian variates would not be zero, corresponding to E in the models shown in Figure 4.1, but would instead be greater than E by about 5 σ units. If the slope of the recovery process decreases with time since the last event (cf. Fig. 4.1), this increase in the "effective" value of E might indirectly lead to modest increase in ensemble precision. The essential conclusion from the preceding calculations, however, is that an ensemble of a million *identical* pacers, triggered by the first event which occurs, ought to behave approximately as an equivalent pacer subjected to a noise process which is about 25% as large as that which affects the single pacer. As the individual oscillators are permitted to depart from the absolute identity in recovery rate assumed here, the precision of ensemble behavior will decrease and eventually approach the precision of the fastest pacer in the ensemble as a lower limit.

Calculations similar to those described above demonstrate that with about 25 pacers in a system of the sort considered above, the variability in system output can be reduced to about half that of the single pacer. *Modest im-*

provement in system precision can thus be readily achieved; but appreciable further improvement in precision requires a massive infusion of additional elements into the ensemble. This kind of result contrasts markedly with that for the pacemaker models defined in Chap. 4, in which variability in system performance is simply proportional to $1/\sqrt{N}$ (Chap. 6).

Chapter 9
General Features of Entrainment: the Type Model

Most experiments designed to examine the synchronization of circadian rhythms have used square-wave light-dark regimes, in which lighting is simply switched on and off at regular intervals. The obvious advantage of this method is its convenience; providing intervening twilights, so as to mimic natural illumination, requires more expensive equipment. An animal's responses to a square-wave light regime might conceivably differ in important ways from those to natural light regimes, but much experimental evidence indicates that simple on-off light cycles are at least *adequate* stimuli to synchronize circadian wake-sleep rhythms. Regardless of whether this process resembles what happens in nature, such observations must be accounted for by any satisfactory model. In this chapter we will therefore first ask how the type model responds to simulated on-off light-dark cycles; gradual changes in intensity will be treated later in the chapter.

In entrainment to a square-wave light regime, diurnal and nocturnal animals differ from each other in several important ways. The first of these is the obvious one, that locomotor activity and wakefulness of a nocturnal animal are primarily confined to the hours of darkness, and those of a diurnal species to the hours of light; but there are other consistent and interesting differences as well, and the two cases will therefore be considered separately.

9.1 Entrainment of Nocturnal Animals: Inhibition by Light

A nocturnal rodent such as the golden hamster, exposed to an on-off light cycle with a 24-h period, ordinarily confines *all* its running-wheel activity to the hours of darkness. Within that limitation, the timing of an entrained nocturnal animal's activity, relative to the lighting, ordinarily assumes one of two extreme positions: (1) if the animal's rhythm has a free-running period (measured in darkness) which is greater than the period of the light cycle, stable entrainment results when the ends of locomotor activity coincide with onset of the lights; in this situation, daily awakening and onset of activity may occur many hours after the lights have been extinguished (Fig. 9.1A); (2) if, on the other hand, the animal's rhythm has a

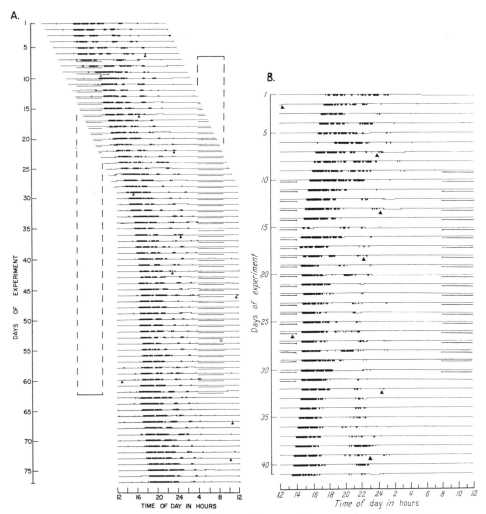

Fig. 9.1 A and B. Actograms showing the entrainment of nocturnal rodents to 24-h light cycles. **A** Entrainment of a hamster, with free-running period greater than 24 h, to a cycle with 5 h of light and 19 h of darkness: the stable phase involved an end of activity which coincided with onset of light (*underlined*); **B** Entrainment of a flying squirrel, with free-running period less than 24 h, to a light cycle with 6 h of light and 18 h of darkness: the stable phase involved an onset of activity about 45 min after the end of light. Note that in part **B**, the preceding free-running period continued essentially unaltered until activity onset occurred within about an hour of the end of the daily lighting, on day 17. Illustrations from DeCoursey (1961, 1964)

free-running period shorter than that of the imposed lighting regime, the entrained phase relationship is such that onsets of wheel running occur very shortly after the lights have been extinguished; the ends of activity may then occur many hours before the lights are again turned on (Fig. 9.1 B).

As a superficial description it appears that light "inhibits wakefulness"; in the one case, that onset of lighting "puts the animal to sleep", thereby speeding up the rhythm; and in the other case, that light "postpones awakening" until extinguished, thereby retarding the rhythm.

If a 24-h light cycle incorporating only a few hours of lighting a day is imposed on a nocturnal rodent and if the timing is such that the light initially does not overlap the animal's activity, then the animal may seem to ignore the light cycle for many days, simply continuing its prior free run until eventually evidence is seen that the light cycle has influenced the rhythm (initial portion of Fig. 9.1B)[8]. This kind of result resembles entrained phasing in that it gives the impression that the animal's locomotor activity must encounter the inhibitory effects of light in order to be affected: that entrainment results from the regular "collision" of wakefulness with either onset or end of the light phase of the cycle.

All these features of the entrainment of nocturnal rodents by square-wave light cycles can be readily reproduced by Coupled Stochastic Systems in their simplest form. To demonstrate that process, simulations were undertaken with the type model, modified by slight variation in $\overline{X}_b$, so as to produce periods slightly longer or shorter than 24 h. The threshold of the discriminator was set at 0.3N for the hours of darkness, and increased to a much higher level[9] during the "light" phase of the entraining cycle, in conformity with the assumption made in Chap. 7 that light raises the threshold for nocturnal animals. The results from two such simulations are illustrated in Figure 9.2; the correspondence with experimental data appears to be satisfactory in all respects. In Figure 9.2A, free-running period of the pacemaker was longer than 24 h, as indicated by both the pre- and post-entrainment data; the entrained phase relationship involved end of activity coincident with (simulated) onset of the lights, and activity onset occurred many hours after the lights went out (cf. Fig. 9.1A). In Figure 9.2B, the free-running period of the pacemaker was shorter than 24 h; at entrainment, activity onset occurred immediately after the (simulated) extinction of the lights, and end of activity occurred long before the onset of lights (cf. Fig. 9.1B); in both cases (particularly evident in Fig. 9.2B), light had no effect on pacemaker behavior until the lighted portion of the daily cycle "collided" with the interval of wakefulness.

The finding that light affects the pacemaker rhythm only at times when the discriminator would otherwise be active follows directly from the stip-

---

8 Particularly striking examples of this phenomenon, using very-short-duration lighting, are illustrated in Figure 3 of Pittendrigh and Daan (1976a).

9 In these simulations, a higher level of threshold (0.6N) was utilized for light than in the simulations of Chap. 7 for constant conditions; this usage is in keeping with the common experimental procedure of using very bright light for entrainment regimes.

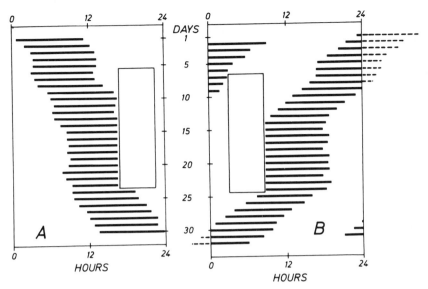

Fig. 9.2 A and B. Actograms showing two cases of entrainment of the pacemaker model by a square-wave threshold regime, with action of light formulated in a manner appropriate for a nocturnal animal (threshold raised greatly, for 6 h, as indicated by *enclosed rectangles,* the equivalent of 6 h of light in a 24-h cycle). $\overline{X}_b = 17.5$ h for part **A**, and 16 h for part **B**; 400 pacers for each simulation; all other parameters as in the type model

ulation that light raises discriminator threshold for nocturnal animals. If at a given time summed pacer discharge of the ensemble has such a small value that the discriminator is inactive, then further temporary increase in discriminator threshold, above that value which prevails in darkness, does not further affect the pacemaker system. The discriminator simply remains inactive, producing no feedback, just as it would if the light had not been experienced. (For example, if, in Fig. 5.3, threshold were to be greatly increased between, say, hours 35 and 40, or 355 and 360, this would be completely irrelevant to pacemaker performance.)

Similar elementary considerations can also readily explain why light cycles can lengthen period of the pacemaker when light occurs near the onset of discriminator activity (cf. Fig. 9.2B): an increased value of discriminator threshold at the expected time of spontaneous onset of feedback will postpone the onset of discriminator activity, by demanding that a larger fraction of the pacer ensemble be discharging simultaneously. A larger value for threshold, to simulate bright light, will be associated with a longer waiting time, just as in the free-running case, in which a higher value of threshold lengthens pacemaker period (Chap. 7). The mechanism by which simulated light can shorten the period of the pacemaker (cf. Fig. 9.2A) is, however, far less transparent. Because of the complexity involved, consideration of this important phenomenon will be postponed to later in this chapter.

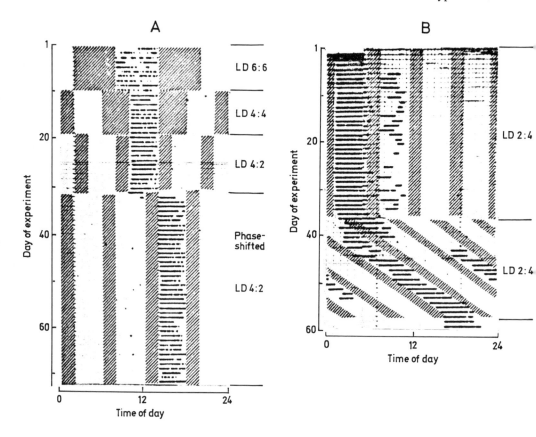

Fig. 9.3 A and B. Actograms showing the entrainment of nocturnal rodents to light cycles with fractional daylengths. A deermouse, in light cycles with periods of 12 h, 8 h, and 6 h, including a phase shift of the 6-h lighting regime; **B** hamster, in 6-h light cycles, as well as in cycles with a period of 6 1/4 h. Times of light indicated by *dark shading*. From Bruce (1961)

Other simulations, equivalent to those of Figure 9.2, have shown that reliable entrainment by a lighting regime with a period shorter than that of the pacemaker's free-running period (example in Fig. 9.2A) can be accomplished only when the free-running period of the pacemaker is relatively close to that of the entraining cycle. If the difference in periods was more than some 30 to 45 min, breakaway from entrainment frequently occurred; and most of the experimental data from nocturnal rodents indicate a similarly restricted range of entrainability. We will return to this point when considering responses of nocturnal rodents to single pulses of light (Chap. 10); stable entrainment to a slightly wider range of stimulus cycles can be attained if certain parameters of the type model are altered in value, but the range remains relatively narrow.

The type model can also duplicate what has been termed frequency de-multiplication (Bruce 1961): synchronization of the activity rhythm of an animal to a 24-h period by square-wave light cycles with periods which are whole fractions of 24 h (e.g., 12, 8 or 6 h). Figure 9.3 presents two experiments with rodents showing this kind of result; Figure 9.4 presents compar-

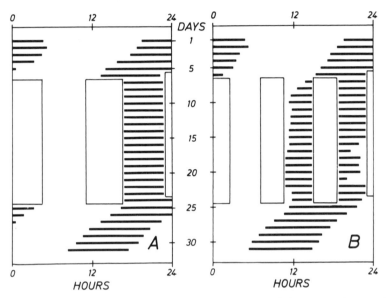

Fig. 9.4 A and B. Actograms showing entrainment of the pacemaker model by square-wave threshold regimes, with nocturnal formulation of threshold, the equivalent of 6 h of light in a 12-h cycle (part **A**), or 4 h of light in an 8-h cycle (part **B**), with times of light indicated by *enclosed rectangles.* $\overline{X}_b = 16$ h, with 400 pacers: all other parameters as in the type model

able results from simulations, with 12-h and 8-h stimulus regimes, for which threshold was again varied stepwise, as for Figure 9.2. The correspondence with experimental data again appears to be satisfactory. The type model can also be readily entrained by a 48-h or a 72-h "light" cycle with a few hours of light per cycle, but in this case, I know of no experimental data from nocturnal animals with which such simulations could be compared.

9.2 Phasic vs Tonic Effects of Light

Experimental data like those shown in Figure 9.1, in which the locomotor activity of a nocturnal rodent is apparently synchronized by its collision with either the onset or the end of the daily light cycle, could easily lead to the impression that light transitions themselves are discrete "signals" for circadian systems. The biological-clock literature makes extensive use of terms like the dawn signal and the dusk signal, the lights-on signal and the lights-off signal. This terminology carries with it the suggestion that a sudden change in lighting conditions may be viewed as the equivalent of a daily alarm clock, which can be used to readjust internal errors in timing. In evaluating the plausibility of such a signalling effect of sudden transitions of light intensity, one can find ready support for the concept in the phasic responses of visual systems. Many neurons are observed to show strong, short-duration bursts of spikes in response to sudden changes in intensity, some to an increase in intensity, others to a decrease; and there are still others which show clear responses to both sorts of sudden change in illumination. The discharges of such neurons might in principle, therefore, act as signals by which a circadian rhythm is synchronized each day by a square-wave light cycle.

An alternative view of the synchronizing effect of lighting, emphasized by Aschoff (1961, and elsewhere), can be derived from consideration of tonic light receptors, which are also widespread in visual systems: cells which discharge continuously, with spike frequency which depends upon prevailing light intensity. In fact, many of the neurons which give strong phasic responses to changes in intensity also show tonic effects under steady illumination, long after their initial burst of activity has subsided.

The circadian literature has avoided the terminology of sensory physiology in considering this distinction. Possible tonic influences of light on circadian systems are instead referred to as proportional or parametric effects; phasic influences, due to sudden transitions in intensity, are referred to as differential or nonparametric effects. Changes in the period of a free-running rhythm due to the intensity of continuous illumination (cf. Chap. 7) are generally thought to be due to tonic − or parametric − effects, since no transitions in intensity have been imposed (although the idea that the animal administers a "light shock" to its visual system when it awakens and opens its eyes, has been suggested as a possible mechanism by which phasic effects might arise, even under constant light conditions). On the other hand, the entrainment of a nocturnal rodent to a square-wave light-dark cycle (cf. Fig. 9.1) could easily be taken to indicate phasic effects of light transitions on the pacemaker, by the manner in which activity onset or end coincides with a step-up or step-down in the lighting regime.

The simulations of Fig. 9.2 demonstrate, however, that phasic effects of light are unnecessary in order to account for this kind of synchronization; the model pacemaker responds only to concurrent lighting conditions, to the momentary, prevailing intensity, in a manner which could be mediated purely by influences of light on tonic receptors. The concepts of a dawn signal and a dusk signal, such as might arise from phasic light receptors, are inapplicable to the simulation results, and are thus unnecessary to account for experimental data like those of Figure 9.1. (This is not to say that phasic effects of light transitions on circadian rhythms do not exist; we shall at a later time be forced to invoke them in a restricted context. The point here is simply that we do not need them *yet.*)

9.3 Entrainment of Diurnal Animals: Stimulation by Light

The synchronization of a diurnal bird by a square-wave light-dark cycle differs in several consistent respects from that of a nocturnal rodent. A light regime can be chosen under which essentially all locomotor activity by the bird is confined to the lighted portion of the cycle — seemingly the simple converse of the general situation with nocturnal animals — but radical departure from that direct temporal correspondence is common with birds. An onset of locomotor activity well before the onset of the lights, such as illustrated in Figure 9.5 is so frequently observed as to be regarded as the usual situation. Furthermore, a diurnal bird will usually remain active

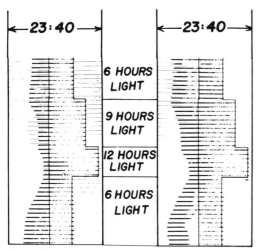

Fig. 9.5. Actograms showing the entrainment of the circadian rhythms of two house finches to light-dark cycles with 23 h 40 min period, incorporating 6 h, 9 h and 12 h of light per cycle. Note that steady-state phasing of the rhythms involved an onset of intense activity several hours before onset of the lights in each case (after Enright 1966). Although most experimental studies of entrainment have used light-dark cycles with a period of 24 h, there is no difference, in principle, in the use of cycles with periods slightly different from 24 h, as in these experiments. An advantage of a non-24-h lighting regime is that an observed entrained phase relationship between animal and light cycle cannot be contaminated by direct responses to some unintentional alternative environmental cycle associated, for example, with daily laboratory noise or the like

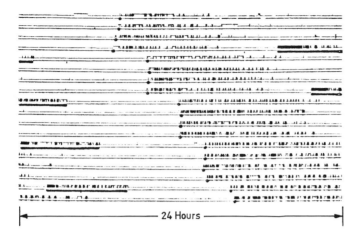

Fig. 9.6. Actogram showing the locomotor activity of a house finch, with a free-running rhythm under constant dim light, except for 6-h pulses of bright light (underlined) administered every 4 days. Note that the bird was intensely active during the hours of supplementary lighting, at times when it otherwise would have been asleep. The conspicuous shifts in timing of the rhythm following each stimulus are a matter considered in Chap. 11. From Hamner and Enright (1967)

for the full duration of the lighted phase of a square-wave light-dark cycle, even when the lights are prolonged to 18 to 20 h out of 24; in contrast, a nocturnal rodent, exposed to the inverse cycle, with 18 to 20 h of darkness, will be active for only 10 to 12 h. In a similar vein, a bird will ordinarily show locomotor activity concurrent with lighting, even when the lights are turned on at an unexpected time, in the middle of its normal interval of sleep (Fig. 9.6). No comparable phenomenon arises when out-of-phase dark pulses are administered to nocturnal rodents. Still another difference is that when a bird is initially placed under a light cycle, the process of entrainment usually begins immediately, in the first cycle, regardless of the phasing between light and the activity pattern; a prolonged continuation of free run in this circumstance, such as often seen in nocturnal rodents (Fig. 9.1B) is not observed.

Thus, the differences between the behavior of a diurnal bird and that of a nocturnal rodent, in response to identical light cycles, go far beyond a simple mirror-image symmetry in the phasing of wakefulness relative to the environmental cycle. At first glance, it might seem quite unlikely that Coupled Stochastic Systems would behave in comparably different manners, depending only upon whether one views the system as applicable to diurnal or nocturnal animals. The assumption was proposed in Chap. 7 that light alters discriminator threshold in opposite directions for diurnal and nocturnal animals; might one not then expect reversed but otherwise comparable responses of the two categories of animals to an imposed light-dark cycle? A reconsideration of Figure 7.8 indicates why this expectation is inappropriate. As that diagram shows, we have assumed that light conditions under which a free-running rhythm is usually observed — darkness for a nocturnal rodent, very dim light (or darkness) for a diurnal bird — represent roughly the same level of discriminator threshold. It is against this background that a light cycle would exert its effect, with light raising threshold, for a nocturnal rodent, to levels not experienced by a diurnal bird, and lowering threshold, for the diurnal bird, to levels not experienced by the rodent. These responses are not, in fact, symmetrical.

The key point, then, is that light is presumed to be much more stimulatory for a diurnal bird, than darkness is for a nocturnal rodent; the evidence leading to that interpretation, derived from experiments under constant conditions, is explained in Chap. 7. With this distinction, the consequences of an imposed light cycle will be qualitatively different for the two classes of animals in several ways; and these differences correspond in detail with differences observed between birds and rodents in the synchronization of their circadian wake-sleep rhythms by on-off light-dark regimes.

Extreme lowering of the discriminator threshold by light for the bird will mean that the discriminator will be activated whenever bright light occurs; even with very few pacers discharging, the summed output would be above threshold. The consequence is that bright light, whenever it occurs, would lead to feedback, capable of accelerating all nondischarging pacers; and, following prior assumptions, would be associated with locomotor activity. This means that locomotor activity will "fill up" the bright phase of any imposed light-dark cycle, and can even spill over into the unlighted phase; and that a light stimulus administered at any time in the pacemaker cycle would be expected to evoke "forced" activity, and thereby conceivably to affect subsequent timing of the system by abnormally timed feedback (cf. Fig. 9.6).

Entrainment of a Coupled Stochastic System is readily achieved by simulating a square-wave light cycle, in which light lowers discriminator threshold in the manner assumed appropriate for a diurnal bird; and, as anticipated in the preceding discussion, the entrainment behavior differs conspicuously from that obtained with the nocturnal-animal formulation of threshold. Two examples of such simulations are illustrated in Figure 9.7, with a threshold value of 0.3N taken to represent darkness (or very dim light); threshold was lowered, during the bright-light phase of the cycle, to 0.1N, a level which would, if given continuously, lead to arrhythmic, continuous activity of the pacemaker system. These simulations show that when the free-running period of the pacemaker was shorter than that of the entraining oscillation, the discriminator began its feedback activity several hours before onset of the lights (Fig. 9.7A). When the period of the pacemaker was longer than that of the imposed cycles in threshold, activity onset co-

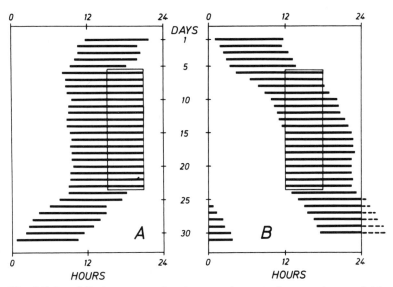

Fig. 9.7 A and B. Actograms showing entrainment of pacemaker model by square-wave threshold regime, analogous to 6 h of light in a 24-h cycle, with lowered threshold during light, as assumed appropriate for a diurnal animal. $\overline{X}_b$ = 16 h and 17.5 h, for parts A and B, respectively; 400 pacers, with all other parameters as in the type model

incided with onset of lighting (Fig. 9.7B). Note, also, that in both cases the simulated light cycle initiated the entrainment process on the first day of application, in contrast to the nocturnal-animal simulations in Figure 9.2.

While these features of the simulations with regard to the phasing and distribution of pacemaker activity are in good agreement with experimental observation, there is one significant aspect of disagreement. In most experimental data for diurnal birds subjected to square-wave light cycles, the daily termination of activity coincides with end of the lighting (cf. Fig. 9.5A; see Eskin 1971, for occasional exceptions). The activity of the pacemaker system in the simulation of Figure 9.7B, however, continued for several hours after lighting had ended. This inconsistency between behavior of a bird and the behavior of the model is no minor matter; a related discrepancy arises in the context of light-pulse experiments (Chap. 11); eventually a modest elaboration of the basic models, invoking phasic effects of sudden light transitions, will be shown to accommodate these effects.

9.4 The "Anomalous" Effects of Threshold on Pacemaker Period

The simulations of Figure 9.2 and 9.7 have demonstrated that cyclic variation in discriminator threshold can entrain Coupled Stochastic Systems, regardless of whether their unentrained period is longer or shorter than that of the imposed cycle. The phenomenon has until now been treated simply as a matter of fact, as an observed aspect of simulations which conforms with experimental data, but in some ways the result is surprising and even counterintuitive when considered in terms of the behavior of individual pacers. In the formulation of the manner in which discriminator feedback affects a single pacer, it was specified that whenever the discriminator is active, the recovery phase of a pacer would be shortened by up to ϵ hours; but feedback from the discriminator can only thereby *shorten* cycle length of a pacer, never increase it. (The basis for this kind of treatment, as explained in Chap. 4, is the assumption that a pacer is a simple relaxation oscillator.) If one, therefore, envisions a single pacer, subject, say, to 12-h blocks of externally imposed feedback in a 24-h cycle, it is reasonably clear that this cycle could synchronize the pacer, so that its onset of discharge roughly coincided with onset of feedback, provided — and this is the critical issue — the basic period of the pacer is longer than 24 h. If the basic period of the pacer were to be shorter than 24 h, there is no way in which it could be synchronized by that feedback cycle. How, then, is the ensemble behavior illustrated in Figure 9.7A to be interpreted, since the free-running period of the pacemaker ensemble was shorter than 24 h?

The source of such counterintuitive results lies in the behavior of the unentrained pacers: those with basic periods somewhat shorter than the period of the pacemaker ensemble. The essence of the phenomenon is the following: the free-running rhythm has a characteristic duration of feedback associated with the resulting period of the pacemaker. If that duration of feedback is artificially prolonged, the period of the pacemaker will be increased because fewer cycles of the unentrained pacers can contribute meaningfully to attainment of threshold.

As was explained in Chap. 5, the unentrained pacers interact with the ensemble rhythm in a systematic way. Their average period is shortened by feedback, and their average phasing is related to that of the discriminator. In the process of gradually scanning through cycles of discriminator activity, their onsets of discharge have a strong tendency to occur during the several-hour interval shortly before onset of discriminator feedback; their discharges therefore contribute regularly and systematically to achievement of threshold during free run. When a light stimulus is simulated which prolongs discriminator feedback in a given cycle, that timing of feedback has the potential of greatly accelerating some of those unentrained pacers, so that their discharges begin at or near end of the bout of feedback which was prolonged by the lights. When that happens, the accelerated unentrained pacers are no longer available for buildup of summed discharge for the next onset of feedback. The subpopulation of pacers leading to the next threshold crossing is reduced; some of its members have been "stolen" by the preceding cycle; and so attainment of threshold will be delayed. Ac-

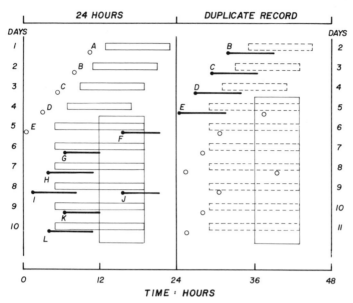

Fig. 9.8. Schematic illustration of the expected behavior of an unentrained pacer with a basic period of 21.5 h, during free run of the pacemaker (pacemaker period of 22 h) and following imposition of a 24-h simulated light regime, where light lowers threshold. Stochastic variation is neglected. Sequential onsets of discharge during free run are indicated by the sequence of points labeled *A, B . . . E,* with the 7-h duration of discharge shown by *solid horizontal bars. Open circles* represent onsets of discharge duplicated in the opposite panel. During the entrainment regime, this pacer had an average period of 18 h. See text for further discussion

celeration of the unentrained pacers can thus lengthen the period of the ensemble.

A schematic example illustrating this process is shown in Figure 9.8. With imposition of the light cycle on day 5, discriminator activity is prolonged. Consider a pacer with basic period shorter than that of the free-running pacemaker, such as the one with discharge shown by heavy solid lines in the illustration. Had the free run simply been continued, that pacer would have contributed to attainment of threshold and to activity onset on days 6 and 7; its onsets of discharge would have occurred at times given by the extrapolation of points A through E. On day 5, however, prolonged feedback causes that pacer to be accelerated to point F, beginning discharge before the end of the lighting. This pacer is therefore no longer available to fulfill its usual role. Some of the unentrained pacers will not have a phase appropriate for this process in the first cycle of the light regime, but all those with phasing similar to that illustrated will be unable to contribute to the buildup to threshold which produces onset of feedback in cycle 6. Attainment of threshold will thus be delayed, and the period of the pacemaker thereby lengthened. This is not just a once-only phenomenon; note that in

Figure 9.8 the same thing also happens in cycles 8 and 9: the 9th onset of
feedback would be delayed, relative to free run, because that same pacer
and others with similar phasing are accelerated in cycle 8. In the entrained
steady state as illustrated, this pacer would contribute to onset of feedback
in two-thirds of the discriminator cycles; its average period, over many cycles, would be 18 h, with four pacer cycles for every three of the discriminator. In free run, its contribution would be available in a far larger fraction
of the cycles, and its average long-run period would be greater.

A closely related phenomenon arises with the nocturnal-animal treatment
of threshold, when simulated light cycles are imposed with a period less than
that of the pacemaker. The essence of that phenomenon is the converse of
that described above: if the duration of feedback is artificially shortened,
the period of the pacemaker will be decreased because more cycles of the
unentrained pacers can contribute meaningfully to attainment of threshold.
When light prematurely terminates feedback, some of the entrained pacers,
which would ordinarily begin their discharge phase near end of feedback,
will not be accelerated. Instead, they will, on the average, wait for the full
duration of ϵ hours before beginning to discharge. Their output can then
"enrich" the subpopulation responsible for the next onset of feedback,
leading to earlier attainment of threshold and a shortening of pacemaker
cycles.

Figure 9.9 offers a schematic illustration of this phenomenon. Part A
shows the free-running situation, in which a given unentrained pacer under-
goes a very short cycle on day 5. Its onset of discharge arises during the
terminal portion of the the 5th block of feedback; its recovery phase has
been shortened in that cycle from about 16 h to about 8 h ($\overline{X}_a = \overline{X}_b - \epsilon$).
Part B shows expected behavior, after imposition of a 24-h light cycle. Be-
cause light terminates feedback earlier than usual on day 5, that block of
feedback cannot accelerate the pacer. Feedback is not available at the time
at which this pacer has recovered for 8 h, so the pacer must proceed with
an unaccelerated cycle, involving a recovery phase of 16 h. The pacer will
therefore be available on day 6 to contribute to attainment of threshold
and onset of feedback. The threshold level of summed pacer discharge will
thereby be achieved sooner than in the free-run case, and the period of the
pacemaker is thus shortened. Note that in this illustration the affected pacer has joined the entrained subset; but other pacers with simewhat shorter
basic period would, granted appropriate phasing, have the potential to accelerate the pacemaker in a comparable way in a fraction of their cycles,
following imposition of the light regime, although they would remain unentrained.

The schematic diagrams of Figures 9.8 and 9.9 resolve the apparent para-
dox of the entrainment simulation: lengthening of the pacemaker period
(diurnal formulation) can be achieved by an exogenously imposed increase

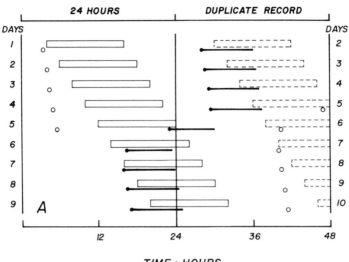

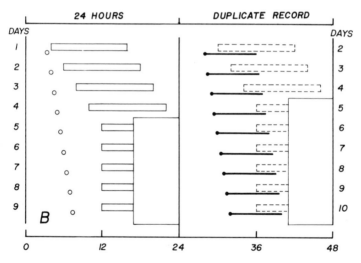

Fig. 9.9 A and B. Schematic illustration of the expected behavior of an unentrained pacer with a basic period of 24.5 h, during free run of the pacemaker (part **A** pacemaker period of 26 h); and following imposition of a 24-h simulated light regime, where light raises threshold (part **B**). Sequential onsets of pacer discharge (stochastic variability neglected) indicated by *solid circles,* with *solid horizontal bars* to show duration of discharge. *Open circles* represent onsets of discharge duplicated in opposite panel. See text for further discussion

in the duration of feedback, thereby radically shortening the cycles of some of the unentrained pacers; shortening the pacemaker period (nocturnal formulation) can be achieved by a light regime which shortens the duration

of feedback, thereby radically lengthening the cycles of some of the unentrained pacers. Thus, the unentrained pacers, which initially may have
seemed to be an unfortunate by-product of the manner in which the loose
system of partial mutual entrainment was achieved, in fact play a very important role in system behavior. They serve to bridge the gap between a
relaxation oscillator – which can only be accelerated by an excitatory stimulus – and the responsiveness of the whole ensemble to light-dark cycles,
in which either acceleration or slowing of the intrinsic rhythm can be
achieved. This is, in my opinion, one of the most interesting points which
emerge from simulation of Coupled Stochastic Systems: an ensemble of
coupled relaxation oscillators can show a kind of behavior which differs
qualitatively from the capacity of any single component and which confers
adaptive versatility to the pacemaker. As we shall see in Chaps. 10 and 11,
this property is to some extent dependent upon the values of certain parameters in the model. Values can be assigned to those parameters which enhance the effects seen in the type model, but other parameter values can
be chosen which prevent this sort of result.

9.5 Entrainment with Gradual Transitions in Light Intensity

As indicated in the introduction to this chapter, experiments utilizing
square-wave lighting regimes have dominated the literature on entrainment
of circadian rhythms. Because of the step-up and step-down transitions in
intensity which accompany those regimes, descriptions of the entrainment
process have often emphasized the concepts of light-on and light-off signals.
Cyclic lighting regimes without discrete transitions in intensity, however,
have not been completely neglected by experimentalists, and the results
obtained have far-reaching implications. The most extensive and interesting
of these studies involve entrainment of the circadian rhythms of several
species of rodents, both nocturnal and diurnal, by quasi-sinusoidal lighting
regimes (Swade and Pittendrigh 1967). The most surprising result is that
light cycles with extremely low amplitude often proved effective. A majority
of the individual animals, regardless of species, were synchronized when the
minimum and maximum light intensities, 12 h apart in a 24-h cycle, differed by as little as a factor of 10; for some individuals, a factor of 2 was
sufficient.

The extremely small rates of change in intensity involved in such regimes
demonstrate clearly that discrete signals, which accompany a sudden change
in intensity, are not essential to circadian entrainment. Phasic light receptors would be of little use to an animal for synchronizing its biological

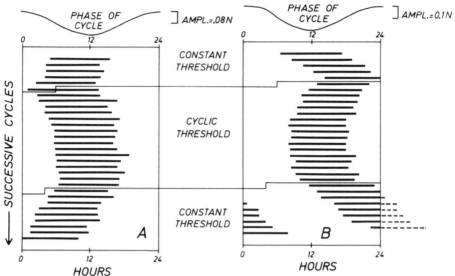

Fig. 9.10 A and B. Actograms showing entrainment of pacemaker model by sinusoidal variations in threshold. Amplitude of the threshold oscillations, as well as their phase, illustrated at the top of each actogram. $\overline{X}_h$ = 16.5 h for part **A**, 18 h for part **B**; all other parameters as in the type model. Note that at entrainment, the stable phase relationship involved activity approximately centered around the low-threshold portion of the threshold oscillation (darkest for a nocturnal animal, or brightest for a diurnal animal), but with greater phase lead in case **A** than in case **B**, reflecting the shorter free-running period in case **A** than in case **B**

clock under such a light regime, unless some very special sort of supplementary assumptions are invoked. Instead, tonic light receptors, the discharge of which varies in a continuous manner with prevailing light levels, are more plausible candidates for mediation of the synchronization, presumably the same sort of receptors which are involved in effects of constant light intensity on pacemaker period (Chap. 7). In general, the ease with which a circadian rhythm can be entrained by quasi-sinusoidal light regimes raises doubts about the importance, for the natural synchronization process, of the sudden light transitions which accompany a square-wave light regime.

We have seen, above, that essentially all aspects of the synchronization of the sleep-wake cycle by square-wave light regimes can be duplicated by the models, as a consequence only of prevailing light intensity, without invoking any phasic receptors. It should come as no surprise, then, that Coupled Stochastic Systems can also account for entrainment by light cycles such as those used by Swade. Two examples of simulations showing this phenomenon are illustrated in Figure 9.10 for which a continuously varying light cycle was mimicked by imposing small-amplitude, 24-h sinus-

oidal oscillations in threshold.[10] In these simulations, the pacemaker did
not immediately synchronize. Instead — and in conformity with the behavior of Swade's animals, which included several species of diurnal and
nocturnal rodents — the pacemaker went through a number of transition
cycles, after start of the simulated light regime, before attaining a relatively
constant phase relationship, and thereafter following the light cycle with
a 24-h period. When the pacemaker was entrained, the feedback activity
of the discriminator occurred (as in the square-wave light regimes) during
the lowered-threshold portion of the imposed cycle: the equivalent of dimmer light for a nocturnal animal and brighter light for a diurnal one, again
in agreement with the experimental data (Swade and Pittendrigh 1967).
The free-running records in Figure 9.10, both prior and subsequent to the
variable-threshold regime, show that the unentrained period of the pacemaker system differed appreciably from 24 h, thereby demonstrating that
true entrainment was involved, and not just chance coincidence of period.
Entrainment of the pacemaker in such simulations has been found to depend, as one might expect, upon the amplitude of the threshold oscillations;
and to require a free-running period of the pacemaker which is not too different from that of the stimulus cycle.

In experiments with rodents exposed to quasi-sinusoidal light cycles, another phenomenon (referred to as oscillatory freerun (!) by Swade and Pittendrigh 1967) has been observed, when the amplitude of the light cycle
was too far reduced. An example of this behavior, for a hamster exposed
to a weak quasi-sinusoidal light cycle, is illustrated in Figure 9.11. A similar phenomenon has long been familiar in other contexts (von Holst 1939).
Aschoff (1965b) and Enright (1965) have used von Holst's term, relative
coordination, for such circadian-rhythm results; the essence of the phenomenon is that when a potentially synchronizing stimulus cycle is sufficiently attenuated in intensity, entrainment does not ensue. Nevertheless, as is
evident in Figure 9.11, the unsynchronized biological oscillation is not truly
free-running; instead, it shows systematic variations in its cycle-to-cycle
period, as it scans various phases of the stimulus cycle. Coupled Stochastic
Systems show entirely equivalent behavior, as illustrated in Figure 9.12.
The varying influence of the stimulus oscillation on the period of the pacemaker is conspicuous; the pacemaker oscillation almost — but not quite —
succeeds in "locking on" to the imposed cycle when discriminator activity
coincides with lowest threshold. Again, the agreement between model and
experimental data appears satisfactory.

---

10 As a sort of calibration for these simulations, it can be seen in Figure 6.9 that a change
    in threshold of 0.1N altered free-running period of the type model by about an hour,
    when provided on a constant basis.

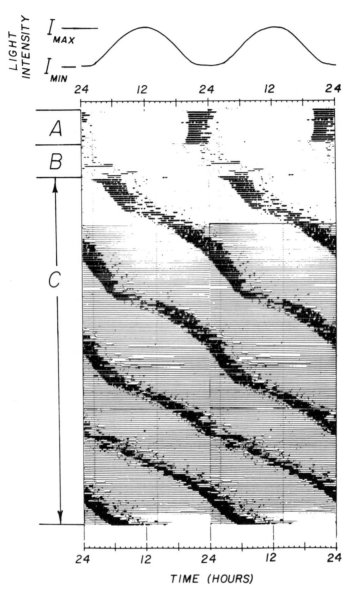

Fig. 9.11. Actogram of a hamster showing relative coordination under fluctuating quasi-sinusoidal lighting regime. For the 14 days in part A, I_{max} was 186 lux and I_{min} was 83 lux. For the 18 days of part B, light intensity was constant (49 lux) and a temperature cycle with $11°$ range ($14.5°-25.5°$ C) was applied. For the six months of part C, the light cycle of part A was applied, but by the end of the experiment, I_{max} had declined to 147 lux and I_{min} to 68 lux. Phase of light regime indicated at top of figure. After Swade and Pittendrigh (1967)

The experimental observations that slow-wave light cycles with small amplitude can entrain circadian rhythms, and that lower-amplitude cycles of this sort result in relative coordination, do not fit conveniently into those

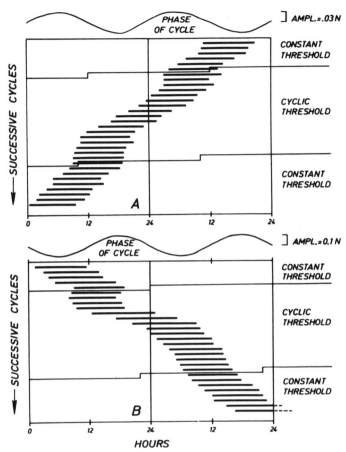

Fig. 9.12 A and B. Actograms showing relative coordination of pacemaker models due to sinusoidal variations in threshold. $\overline{X}_b$ = 16 and 18 h for parts **A** and **B**, respectively. Other details as in Figure 9.10

interpretations for the action of light on circadian systems which are based on signals at dawn and dusk. Such results have therefore led to a supplementary and rather complex explanatory scheme, the "velocity response curve" (Swade and Pittendrigh 1967; Swade 1969; Daan and Pittendrigh 1976b), which postulates that the effects of single, brief pulses of light upon a rhythm can be integrated over the entire cycle, and then multiplied by a transforming factor, to account for responsiveness to slow-wave light cycles. It is of more than incidental interest, therefore, that such experimental findings can be accounted for so readily and naturally in the present context. The single assumption that concurrent light intensity determines discriminator threshold in Coupled Stochastic Systems provides a unitary basis for interpreting three broad classes of experimental data: effects of constant light intensity, effects of square-wave light cycles, and effects of slow-wave sinusoidal light regimes.

Other simulations have shown that entrainability by slow-wave oscillations in threshold is a very general property of the models. Results like those of Figure 9.10 (and Fig. 9.11, when amplitude is sufficiently reduced) have also been obtained in simulations in which α, β, $\overline{\delta}$ and ϵ were assigned values quite different from those of the type model. It is of course important for evaluating the models as a class to note that they can so readily reproduce entrainment by light regimes which slowly change in intensity, the more so as the natural 24-h light-dark cycle involves only gradual changes in light intensity, during twilights, rather than instantaneous steps of intensity like those of a square-wave light regime. There is, however, an unfortunate corollary: entrainability by such stimulus regimes, as simulated by slow cycles in discriminator threshold, is such a general property of the entire class of models, compatible with a broad range of possible parameter values, that the available experimental data do not serve as a meaningful guide for the selection of a particular model of the class. The results do not serve usefully to restrict the choice of parameter values. Instances in which a particular experimental observation can be reproduced by only a small subclass of models (such as are considered in subsequent chapters) are more valuable, since restriction of acceptable parameter values leads to predictions which are quantitatively constrained.

Chapter 10
Responses to Single Light Pulses. Part I: Nocturnal Rodents

The responses of an oscillatory system to single perturbations can often guide one to a deeper understanding of the system's dynamics. Such experimentation has been emphasized in several studies of the circadian wake-sleep rhythm. The issue examined in this chapter, as well as the next, is the extent to which Coupled Stochastic Systems can accomodate the data obtained from treatment of nocturnal rodents, and of diurnal birds, with single light pulses.

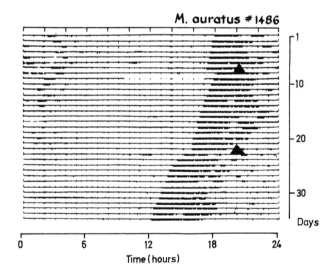

Fig. 10.1. Actogram of a hamster under free-running conditions, with the imposition of two 15-min light stimuli, as indicated by *triangles*. Note that the phase delay, in response to the first stimulus, was completed immediately, but that phase advance was only gradually attained, over several circadian cycles. From Daan and Pittendrigh (1976a)

In a typical experiment of this sort, the animal is maintained under constant conditions long enough to document the pretreatment properties of a steady-state, free-running rhythm in locomotor activity; a light pulse of some chosen intensity and duration is administered; and the changes induced in temporal pattern of the rhythm are then observed. The experimental design is simple enough, but the results can be complex. Figure 10.1 presents an example in which two 15-min light pulses were administered to a hamster (*Mesocricetus auratus*). Following the first treatment, the animal's rhythm was delayed by about an hour, with the entire delay being evident on the first day after treatment. As a result of the second treatment, the animal's rhythm was advanced by the stimulus, but at least three circadian cycles were required before the rhythm returned to steady state. During that entire interval, the overt period of the rhythm was somewhat shorter than during free run. Qualitatively similar changes in timing are shown in Figure 10.2, from an experiment in which a hamster was treated with 12-h light stimuli.

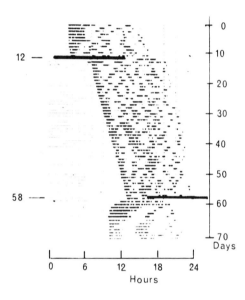

Fig. 10.2. Actogram of a hamster under free-running conditions, with the imposition of two 12-h light stimuli, as indicated by *solid bars*. Note that phase delay was completed within 2 cycles, but that phase advance involved 4 to 6 transient cycles before steady-state period was attained. From Pittendrigh (1965)

A first and central conclusion which follows from experiments of this sort is that the amount, and even the direction by which timing of a circadian rhythm is altered (usually called a shift of phase: one or more cycles during which instantaneous period differs from its free-running value), depend upon the timing within the circadian cycle at which the stimulus is administered. The effects of long-duration stimuli like those shown in Figure 10.2 , however, are not easy to interpret in that regard. Note that the two 12-h treatments overlapped extensively in their timing relative to the animal's wake-sleep rhythm; in both instances, the animal experienced light during the interval from about 3 h to 9 h after activity onset. The observed difference in results (phase delay in one case, phase advance in the other) might conceivably be due only to those portions of the stimuli which differed in the two cases, but one cannot simply assume that; nonlinear interactions between delaying and advancing effects might also exist.

To avoid such complications, most experimentation on shifting of circadian rhythms by single light pulses has emphasized use of short-duration stimuli, like those involved in the experiment of Figure 10.1, even though such stimuli are conspicuously unecological, and even though the basic phenomenon to be understood involves natural light stimuli with a duration of many hours. The results of each single-pulse treatment constitute a measure of the amount of phase shift (delay or advance), as a function of the time (or phase) within the circadian cycle at which the stimulus was administered. An array of such single-pulse results can be plotted, with phase shift as ordinate and time of treatment as abscissa, leading to what is now generally known as a phase response curve. Several examples of phase response curves obtained for nocturnal rodents are presented in Figure 10.3.

A troublesome element in this method of data summary arises because of the finding that phase advances of a circadian rhythm often require several days for completion (evident with the second treatments in Figs. 10.1 and 10.2). Some publications have reported only the amount of phase shift measurable on the first day after treatment (about 90 min advance in Fig. 10.2); others have presented only data on the net difference in timing between the pretreatment rhythm and the steady state observed after so-called transients have subsided (a total of more than 3 h advance in the lower part of Fig. 10.2). Very rarely are both values given. Some of the curves of Figure 10.3 (as indicated in the figure legend) are based on one method of measurement and others on the alternative.

10.1 Phase Shifts of Coupled Stochastic Systems

Simulations with the type model were undertaken in which single short-duration increases in threshold were applied, formulated so as to represent 2-h pulses of bright light administered to a nocturnal animal. The results are presented in Figure 10.4 in terms of phase shift measured in the first cycle following treatment, as well as in the two subsequent cycles. Correspondence between simulation results and the experimental data is by no means complete, but it is worthwhile, at this point, to note several significant aspects of agreement: (1) stimuli during the inactive portion of the pacemaker cycle produced no phase shift whatever; (2) stimuli at or shortly after onset of activity resulted in a delay of the subsequent rhythm; and (3) stimuli in the latter part of the activity time led to advance of the subsequent rhythm.

The lack of response by the pacemaker models to threshold-raising stimuli during the inactive portion of its cycle has already been alluded to in

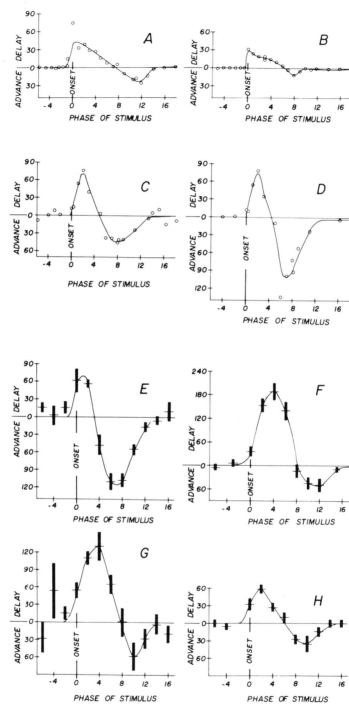

PHASE OF STIMULUS: HOURS AFTER ACTIVITY ONSET

Fig. 10.3 A–H.

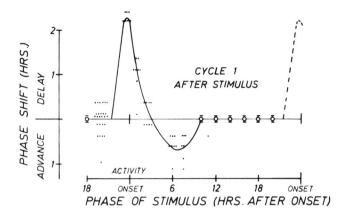

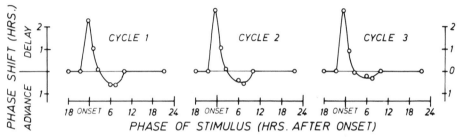

Fig. 10.4. Phase-shifting results for the type model (with 800 pacers), in response to 2-h increases in threshold. *Small points* in upper graph: replicate measurements of phase shift, obtained during first cycle after treatment; *open circles with vertical bars* in upper graph: mean, and standard error, for the 26 simulations with stimuli at hour 20; the lack of shift at hours 10 to 20 was inferred on the basis of the understanding of how the stimuli at this time would influence the system (see text), and was not directly measured in simulations. *Lower graphs:* mean values of phase shift from cycle 1 (i.e., means of data sets in upper graph), as well as in cycles 2 and 3 following stimuli. Note that although phase delays were stable, the magnitude of phase advances tended to decrease slightly in cycles 2 and 3

◀──

Fig. 10.3 A–H. Phase-response curves obtained from administration of single brief light pulses (10 to 15 min) to nocturnal rodents. **A** and **B** data from two individual flying squirrels, *Glaucomys volans,* as measured on the first day after stimulation (from DeCoursey 1960); **C** and **D** data from a single hamster, *Mesocricetus auratus,* as measured on the first day after stimulation (**C**) and at eventual steady state (**D**) (from DeCoursey 1964); **E–H** data from four species, with 5 to 13 individuals of each species, with individual observations grouped by 2- to 3-h bin widths (from Daan and Pittendrigh 1976a); *horizontal lines* represent mean phase shift, *vertical bars* represent ± one standard error; **E** hamster, *M. auratus;* **F** mouse, *Mus musculus;* **G** deermouse, *Peromyscus maniculatus;* **H** deermouse, *P. leucopus.* Note that part **F** has a reduced vertical scale. Data in parts **E** to **H** are only approximations, based on my conversion of "circadian time" to hours after activity onset and of phase shift in "angular degrees" to shift in minutes

Chap. 9; the explanation is that if summed pacer discharge at a given time is so small as to be less than the threshold which prevails during free run in darkness (meaning that there is no feedback from the discriminator), further increase in threshold due to light cannot alter behavior of the ensemble. A comparable and seemingly complete lack of responsiveness to light stimuli during their normal sleep-time is characteristic of all species of nocturnal rodents which have been investigated (Fig. 10.3). The experimental observation might, of course, simply mean that a sleeping mammal is unaware of the light stimulus, but DeCoursey (1961a) demonstrated the inadequacy of that explanation by waking squirrels in the midst of their sleep, keeping them awake by forcing them to activity in running wheel, and then administering the light stimuli. No phase shift of the rhythm resulted. The present models can, of course, adequately account for this lack of response, without requiring that the stimulus be unseen. The pacemaker itself is unaffected by visual input at this phase, so whether the animal has its eyes open or not is irrelevant here.

Phase delay of the pacemaker model due to light stimuli just before onset of activity can be readily understood by analogy with the effects of continuous bright light. An increase in threshold, at the time when feedback would ordinarily be initiated, requires greater summed pacer discharge; onset of feedback is postponed because of an increase in waiting time for this greater sum to be attained by spontaneously discharging pacers. The resulting delay of activity onset is propagated as a permanent shift of the pacemaker rhythm because the feedback-induced resynchronization of recovering pacers, which is an essential determinant of phase for the next cycle, has been delayed.

Stimulation of the pacemaker system shortly *after* activity onset produces phase delay by a similar process, because the feedback-induced resynchronization of entrained pacers after activity onset is not an instantaneous event, but instead requires several hours after onset of feedback (see Figs. 5.3 and 5.6). Therefore, when threshold is raised and feedback is temporarily interrupted early during the activity time, many of the entrained pacers which would have begun to discharge due to feedback at that time will have their discharges postponed. A delay of this fraction of the entrained pacers will contribute in the next cycle to a phase delay of the entire system because some of the affected pacers would have, if left undisturbed, contributed to the next onset of activity; because of their delay due to the stimulus, they are less likely to do so. Phase delay, then, whether due to stimulation just before or somewhat after activity onset, represents an influence on the synchronized pacers, a simple postponement of some of their onsets of discharge, which alters subsequent timing of the pacemaker system.

Phase advance of the pacemaker model, due to an increase in threshold late in the activity time, is a more complex process, mediated by the unsynchronized pacers which have "basic" periods shorter than that of the pacemaker ensemble. The process has already been considered in elementary form in Chap. 9. Under free-running conditions, these pacers often begin their discharge phases near the end of discriminator activity (Fig. 5.5). When feedback is interrupted late during the activity time, the appropriately phased unsynchronized pacers will fail to be accelerated in that cycle; without feedback, their onset of discharge will, on the average, be postponed by an amount equal to ϵ (Excitation, i.e., 8 h in the type model), and their discharge can then contribute to the ensuing onset of discriminator activity, leading to an earlier attainment of threshold in the next cycle, as illustrated in Fig. 9.9. Phase delay of this group of unsynchronized pacers permits them to join the subset of pacers which produce the next activity onset. The resulting phase advance of the system will thereafter be propagated primarily by the entrained pacers, which have been thereby reset to an earlier time.

In order for phase advance to be achieved by this indirect mechanism, the discharge of the delayed, short-period pacers must persist through (and thereby contribute to) the next onset of discriminator feedback. For this to occur, those pacers which would ordinarily begin discharge at the time of stimulation, and which are delayed by ϵ hours, must have a duration of discharge, Y_b, sufficiently long that the sum of ϵ and Y_b is at least as great as the normal duration of discriminator inactivity (Fig. 10.5). This requirement is fulfilled by the type model, but just barely so: duration of pacemaker feedback is about 10 h, meaning that inactivity of the discriminator lasts about 14 h; the unentrained pacers have basic periods of somewhat less than 24 h, which, with $\bar{\delta}$ (the Discharge Coefficient) of 0.5, means that the typical value for Y_b is somewhat less than 8 h. With ϵ of 8 h, the sum of Y_b and ϵ will on the average be somewhat greater than the duration of inactivity, but only slightly so. The type model is therefore near the limits of parameter values which can permit phase advance to occur by this process. Beyond this (as can be seen in Fig. 5.5), only a relatively small fraction of the cycles of unentrained pacers in the type model have onset of discharge at a time appropriate for the advance, so the magnitude of the expected effect will be small.

With this situation in mind, let us reexamine the phase-shifting results shown in Figure 10.4. The maximum phase advance obtained in the simulations was only about 30 min. In contrast, several sets of experimental data from nocturnal animals (Fig. 10.3) indicate that a single light pulse can induce phase advance of at least an hour, and often far more. Another reflection of this inadequacy of the type model, referred to in Chap. 9, is that entrainment of the pacemaker to light-dark cycles with period shorter than

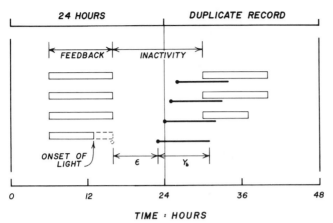

Fig. 10.5. Schematic illustration of the requirement that $\epsilon + Y_b$ must exceed the duration of inactivity by the discriminator. Times of feedback indicated by the elongate *open rectangles;* successive discharges by a pacer with critical phasing indicated by *heavy horizontal lines.* In the absence of the light stimulus, this pacer would have begun its fourth discharge at the end of feedback, as shown by the *small dotted circle.* Since the light stimulus prematurely terminated feedback, this fourth onset of discharge is postponed by ϵ hours. Only if discharge continues on through the expected time for next onset of feedback can this pacer accelerate the system

that of the free-running rhythm (and therefore requiring phase advance of the pacemaker, cf. Fig. 9.2A) was not satisfactorily stable unless the difference between free-running and entraining periods was less than about 45 min.

The fact that the pacemaker system, in its type-model form, shows such weaknesses is a valuable piece of information. An even more important conclusion can be drawn from the preceding considerations: if values appreciably smaller than those in the type model were to be assigned to the Discharge Coefficient, $\bar{\delta}$, and to Excitation, ϵ, the pacemaker system would not even be capable of phase advance by this mechanism at all. It is worth emphasis here that experimental results for which a model *cannot* account are the ones most likely to enlarge our understanding of the real pacemakers. Unless supplementary assumptions are invoked, Coupled Stochastic Systems, in which only small values are assigned to $\bar{\delta}$ and ϵ, are apparently excluded as possible forms for the pacemaker of nocturnal rodents in which light stimuli can evoke phase advance. The next question which arises is whether other members of the class of models are more satisfactory than the type model in terms of the achievable phase advance.

A first possibility, which might enhance the capacity of the model to achieve phase advances, would be to increase the value of threshold assigned for free-running conditions; a larger fraction of the pacers thereby become unentrained, and it is those which have the potential of inducing phase ad-

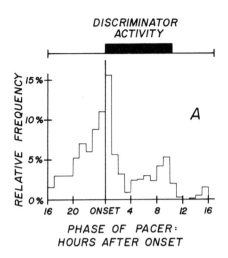

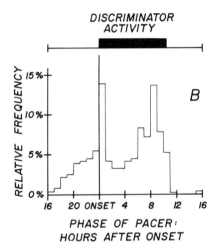

Fig. 10.6 A and B. Timing of onset of discharge by unsynchronized pacers, relative to discriminator feedback activity. A type model; B modification of type model, in which $\epsilon = 12$ h. Each of about 50 pacers contributed at least 20 values to the histograms, from 500-h simulations

vance. Another consequence of that modification, however, is that the duration of feedback activity would be reduced (cf. Fig. 6.9), so that the re-phased pacers would be somewhat less likely to contribute to the next onset of feedback. As can be seen from a consideration of Figure 10.5, that residual problem might be remedied by increasing the value of the Discharge Coefficient, $\bar{\delta}$ (with a concommitant decrease in the Mean Duration of Recovery, $\overline{X_b}$), or by increasing the value of Excitation, ϵ, over its type-model value. Simulations have demonstrated, however, that increasing the value of ϵ alone, with no change in the value of threshold or in $\bar{\delta}$, is sufficient to accomplish the desired objective of increasing the magnitude of achievable phase advance. Stable and reliable advances of the pacemaker, on the order of 2 h, were induced by stimuli late in the activity time, when Excitation, ϵ, was increased from 8 to 12 h.

Two factors contribute to this result: with a large value for ϵ, unsynchronized pacers, which are delayed by the stimulus, will contribute consistently to the next onset of discriminator activity even if Y_b, the duration of pacer discharge, is rather brief (cf. Fig. 10.5); and, even more important, a larger fraction of the cycles of the unentrained pacers will be phased such that their onset of discharge will occur near the end of feedback activity. Figure 10.6 shows the contrast in the phasing diagrams for the unentrained pacers in the type model (part A), and in a model with ϵ of 12 h (part B). When ϵ is large, the unentrained pacers will thereby have a greater probability of participating in the phase-advance process. This modification of the type model also has a salutary effect upon the entrainability of the pacemaker to light-dark cycles with period shorter than the free-running period of the system (cf. Fig. 9.2A). Increasing Excitation, ϵ, from 8 to 12 h permits acceleratory entrainment of the pacemaker to periods as much as 2 h shorter than the free-running value, instead of some 30 to 45 min, as with the type model.

10.2 Transients During Phase Advance: Their Origin and Significance

In simulations based on this modification of the type model, it was often observed that achievement of a phase advance required two circadian cycles for its full expression, although phase delay was generally completed within a single cycle. This result resembles, at least qualitatively, the striking difference in the manner in which phase advances and phase delays are typically achieved by nocturnal rodents (cf. Figs. 10.1 and 10.2). While such results were not obtained in every test simulation of that sort, and while there were fewer transients than are commonly observed in the experimental data, the result is of sufficient interest and potential importance to deserve further attention. How might such transient cycles arise in the pacemaker system? A detailed consideration of this question is provided in Appendix 10.A. It is shown there that such transients are to be expected in connection with phase advance of Coupled Stochastic Systems, but not with phase delay; and that modest changes in parameter values of the model can enhance these transients, so that they persist for as long as four or five cycles, as is commonly seen in the experimental data (cf. Figs. 10.1 and 10.2).

10.3 Interspecies Variations in Phase-response Curves

The data summarized in Figure 10.3 indicate several differences among rodent species in the phase-response curves obtained from similar experimental protocols: differences in the amount of phase delay and of phase advance, as well as modest differences in the phase within the cycle which separates advancing responses from delays. DeCoursey's data on flying squirrels even document consistent differences in responses of different individuals of a single species. I have made no attempt to select parameters for specific models, so as to reproduce this kind of detail. Preceding considerations, however, suggest some of the ways in which that might be undertaken.

If one seeks a model which will show large phase advances (e.g., for the hamster), it is necessary to have many unentrained pacers in the system, meaning that the value assigned to threshold under free-running conditions should be relatively large; the greater the threshold, the larger the fraction of the ensemble which is unentrained. Large phase advance also requires that the unentrained pacers begin often to discharge, in the free-running situation, at or near the ends of discriminator feedback. We have seen (Fig. 10.6) that a large value of ϵ (Excitation) can contribute to meeting this requirement. For the case in which the organism shows many transients before single-pulse phase advance is completed, it is shown in Appendix 10.A that a small value for Between-pacer Variability, β, is appropriate; reducing the value of γ (Discharge Variability) would also enhance transients, by similarly reducing the variance of basic periods in the ensemble.

Phase delay due to a stimulus shortly after activity onset is envisioned here as being due to the delay of entrained pacers (as well as some few of the unentrained pacers): elements which sometimes begin their discharge well after onset of feedback, and which nevertheless would have a significant chance of contributing to the next subsequent onset of activity. In models of the general class considered in this chapter, entrained pacers cannot show such behavior unless the Stochastic Coefficient, α, is relatively large; a small value of α, granted that ϵ is large, will tend to keep such pacers in a phasing quite close to onset of discriminator activity, so that phase delay would arise only due to stimuli very early during activity time.

In brief, then, none of the differences among the phase-response curves of nocturnal rodents illustrated in Figure 10.3 appears to me to be inexplicable in terms of a model of the general sort considered here, granted appropriate choices of parameter values, except for one essential aspect of the experiments, described below.

10.4 Failure of the Elementary Class of Models

In the consideration of the responses of nocturnal rodents to single pulses of light, I have until now tacitly ignored a critical way in which the models cannot reproduce a general feature of the experimental data. The maximum attainable phase delay of the pacemaker does not appreciably exceed the duration of the light pulse used, and it will usually be somewhat less.[11] Experimental studies of nocturnal rodents have shown, however, that a light pulse of 10 or 15 min duration can induce phase delay of an hour or more (Fig. 10.3). I have found no way to remedy this discrepancy between model and experiment by other choices of parameter values; it is, I think, an inescapable consequence of the assumption that discriminator threshold is determined only by the prevailing, concurrent light intensity, so that a 10-min stimulus can produce only a 10-min effect on threshold. No such equivalent problem would arise in the case of stimuli which lower threshold; a 10-min stimulus, which induces feedback at an unexpected time, could accelerate the discharge of pacers by an amount as great as ϵ hours, but the converse, with stimuli which briefly raise threshold, cannot occur.

If one carefully examines the activity recordings of rodents kept for many days in complete darkness and then subjected to a single brief light stimulus during their wheel-running activity, it is common to note that treatment of a few minutes is associated with an interruption of the animal's wheel-running for an hour or more (e.g., DeCoursey 1964, Fig. 1A; note also that in Fig. 10.1, the light stimulus on day 7 completely terminated locomotor activity for that entire day). In these cases, the animal's behavior demonstrates that a brief light pulse has an inhibitory effect on wheel-running activity which persists far longer than the stimulus itself. Because locomotor activity has been taken here as an index of the status of the discriminator, this phenomenon in itself is enough to call into question the assumption that light alters discriminator threshold only on a moment-to-moment basis. Both the interruptions of locomotor activity and the magnitude of the phase shift resulting from brief light pulses can be accommodated by the supplementary assumption that the sudden onset of bright light (which has until now been assumed only to raise prevailing discriminator threshold to a higher level) is accompanied by a transient overshoot in the response of threshold; that threshold is initially increased to a level

---

11 Certain simulations illustrated later show modest exceptions to this expectation, but those exceptions are sufficiently small in magnitude for the problem addressed here to remain.

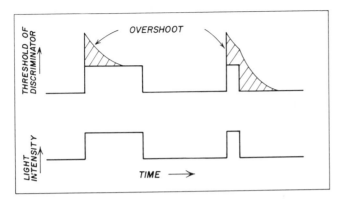

Fig. 10.7. Schematic diagram of the overshoot of threshold, postulated to be associated with sudden increases in light intensity. It is assumed that recovery from this overshoot, to that value of threshold dictated by prevailing light intensity, proceeds at an exponential rate with a half-time on the order of an hour or two

higher than that prevailing under constant light, and that this supplementary effect then decays gradually over several hours. Figure 10.7 outlines this proposal in schematic form. In terms of the physiological mechanism which might underlie such an effect, any of several kinds of adaptation phenomena, central or peripheral, could provide an adequate explanation. For the sake of symmetry, one might well also entertain the idea of a similar overshoot (actually, an undershoot) of threshold, following sudden extinction of the lights, but there is nothing in the experimental data which demands that effect; and some sorts of accommodative processes, such as retinal bleaching and dark adaptation, would not be expected to be symmetrical in this way.

A detailed quantitative formulation of this overshoot of threshold, due to sudden onset of lights, would require at least two parameters, one for the magnitude of the initial effect, and another for its rate of decay. These parameters might be evaluated by phase-shifting experiments in which *two* light pulses are administered, separated by variable intervals (from, say, zero to 4 h); but such experiments have apparently not been undertaken. In the absence of such data, I have simply assumed that a brief light pulse can be simulated by an elevated value of threshold for 2 h: that a 10-min light pulse can interrupt discriminator feedback for 2 h. It is easily imaginable that a response of this sort might depend upon a variety of factors, including the magnitude of the step-up of intensity, the animal's concurrent state of arousal, and the extent of prior dark adaptation. Hence, the proper quantification of this phenomenon might well complicate the pacemaker model to a far greater extent that the sketch of Fig. 10.7 recognizes.

Note that this overshoot assumption will not play any detectable role in a light cycle which includes twilight transitions, nor in any square-wave light cycle where bright light is more than an hour or two in duration. Despite the limited applicability of this process to normal entrainment, it would be inappropriate to gloss over the broader implications of making such an assumption. Here, for the first time, existing experimental data have demanded fundamental elaboration of the general class of models, a modification which goes far beyond simply altering the value of one or another of the initial parameters. Up to this point, the only influence of light upon the pacemaker was assumed to be a tonic effect, with discriminator threshold directly responsive to prevailing, concurrent intensity; here we have acknowledged the necessity of also incorporating phasic influences of sudden light transitions. In the next chapter, we will see that this same sort of elaboration of the models is also essential in order to accommodate certain of the data for diurnal birds.

Appendix 10.A. Transients Following Phase Shifts

The key to an understanding of the origin of transients following phase advance of a Coupled Stochastic System lies in the fact that some of the pacers which contribute to phase advance have basic periods only very slightly shorter than the free-running period of the pacemaker system. The diagrams of Figure 10.8 offer two schematic descriptions of the expected behavior of unentrained pacers, following a threshold-increasing light pulse late in the activity time, with α, the Stochastic Coefficient, set to zero, so as to portray the typical time-average behavior. In part A, the expected cycles of a series of pacers, each with a basic period of 22 h, are shown with a different symbol for each of several pacer phases, relative to discriminator activity. The onsets of discharge of the pacer with phasing appropriate to produce advance of the pacemaker are indicated by close circles. In cycle 5, that pacer would not be accelerated, because of the exogenously imposed shortening of feedback, and it would therefore contribute to an earlier onset of feedback in cycle 6. Since it would, however, be accelerated in its next onset of discharge, to the end of the sixth block of feedback, it would not contribute to transients. Any residual acceleration of the onset of feedback in cycle 7, beyond the free-run value, must have other origins.

In part B of Figure 10.8, dealing with pacers with basic period of 23.4 h, the situation is different and more interesting. The pacer with "critical phase" for producing phase advance (again shown by closed circles) can be expected to contribute to an earlier onset of feedback not only in cycle 6, but also in cycles 7, 8, and 9. Instead of beginning its discharge at the end

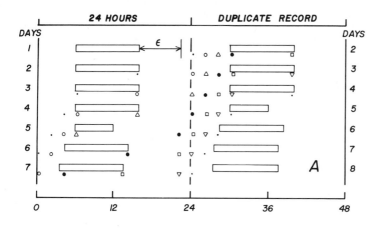

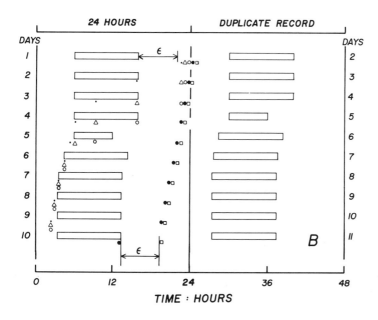

Fig. 10.8 A and B. Schematic illustration of expected onsets of discharge by unsynchronized, short-period pacers in successive cycles, with the system subjected to a single light pulse on day 5 which prematurely terminates feedback. **A** Expected behavior of five pacers, each having a basic period of 22 h, with each pacer separated in phasing by 2 h, as indicated by different symbols. The pacer with critical phasing to produce phase advance of the system is indicated by the large *solid dots*. **B** Similar graph for five pacers which have a basic period of 23.4 h, initially separated by 1/2 h in phasing. The critically phased pacer (*solid dots*) would contribute to additional acceleration of the system for several cycles after the initial phase shift. See text for further discussion

of the fifth block of feedback — as it would have done in the absence of stimulation — this pacer would require the full time until cycle 10 to reach this expected status. The distinction between parts A and B of Figure 10.8 can be summarized as follows: designating the free-running period of the pacemaker system as τ_d and the amount of phase advance on the *first* day after treatment as $\Delta\phi$, any pacer with basic period between τ_d and $\tau_d - \Delta\phi$ will be able to contribute to subsequent further phase advance; it can give rise to transient cycles.

With this understanding of the mechanism by which transients arise in a Coupled Stochastic System, it becomes a simple matter to amplify the process; the requirement is for more pacers with basic period relatively close to (but less than) that of the pacemaker, and that can be readily achieved by decreasing the inter-pacer variability in basic period, i.e., by reducing the magnitude of β, Between-pacer Variability. Simulations which incorporated this further modification of the type model ($\beta = 1$ h rather than 3 h) consistently gave rise to several transients in the process of attaining phase advance. Examples are shown in Figure 10.9, parts C and D. Note that while several cycles were required for phase advance to be completed, the phase-delay process saturated within a single cycle (parts E and F of Fig. 10.9).

This major difference in the manner in which phase advance and phase delay proceed, given a large value for ϵ (Excitation) and a small value for β (Between-pacer Variability), is in excellent agreement with extensive experimental data on nocturnal rodents, of which the cases shown in Figures 10.1 and 10.2 are representative. The phenomenon follows as a natural consequence of the fact that phase delay of the pacemaker model is primarily due to lengthening the cycles of entrained pacers, and that phase advance is due — at least initially — to an effect upon unentrained pacers. Since the occurrence of several transient cycles can arise in the models only due to unentrained pacers with basic periods quite close to that of the free-running pacemaker, the observation of a series of transients in experimental data can be interpreted as setting an upper limit on the acceptable variability in basic period among the pacers of the ensemble.

The existence of transient cycles following phase advance of the pacemaker models is of primary interest here because of its detailed correspondence with a previously unexplained aspect of experimental data. It is also noteworthy, however, because it constitutes a property of the pacemaker

———————————————————————————————▶

Fig. 10.9 A–F. Simulations with a modification of the type model, in which ϵ was increased from 8 h to 12 h, and β was reduced from 3 h to 1 h, with 800 pacers in the ensemble. Parts A and B are replicate, 25-cycle free runs of the pacemaker; for parts C to F, 2-h stimulations with an increase in threshold were applied, as indicated by *open rectangles*. All calculations made using a single ensemble of pacers, with identical parameters, and an identical sequence of values of z_j up to hour 100. Note the several transient cycles following stimulation in parts C and D

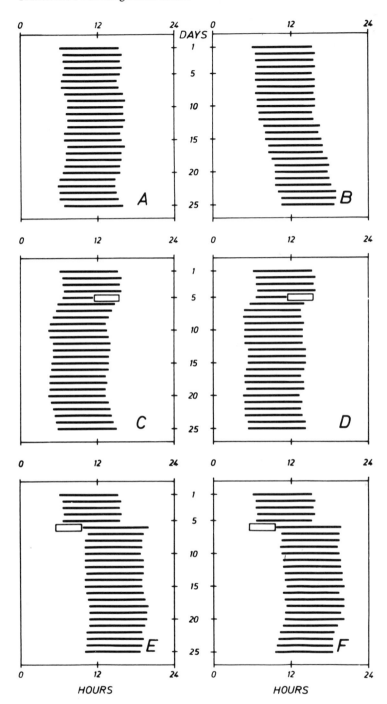

Fig. 10.9 A—F.

ensemble which is not found in the behavior of the individual elements of which the ensemble is composed, when they are acting in isolation; it is a consequence of the coupling. Each of the pacers has been formulated so as to behave as an elementary sort of relaxation oscillator. A single pacer therefore has no intrinsic capacity to show transients; each of its sequential cycles is completely independent, in terms of all internal processes which determine its behavior. The discriminator which couples the pacers cannot, of course, be directly responsible for transients since its behavior has been formulated as a purely passive element, responding only to its concurrent level of input. Nevertheless, the system as a whole, during its transients, gives evidence of "remembering" its prior state over several circadian cycles. We shall, in Chap. 12, see similar phenomena of an even more dramatic sort.

Chapter 11
Responses to Single Light Pulses. Part II: Diurnal Birds

A series of phase-shifting experiments based on single-pulse treatment of the circadian rhythms of diurnal birds has been published by Eskin (1971); he administered 6-h light stimuli to house sparrows. His results are graphed in Figure 11.1. It can be seen there that the nature of the birds' responsiveness to single light stimuli differs in several conspicuous ways from that of nocturnal rodents. Phase shifts of up to 8 h were obtained, compared with

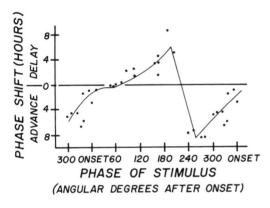

Fig. 11.1. Phase-response curve for the house sparrow, subjected to 6-h light stimuli under otherwise constant dark conditions. Phase of stimulus (onset of light relative to onset of activity) expressed in angular degrees; since values for free-running period were on the order of 25 h, 60° is equal to about 4.2 h. Values derived from Eskin (1971)

a maximum of 3 to 4 h in the most sensitive of rodents [12] (Fig. 10.3). These large phase shifts of the sparrow's rhythm were obtained when light fell in the midst of the birds' normally inactive time, some 14 to 18 h after activity onset — a timing at which a nocturnal rodent is completely unresponsive to light. With the birds, stimuli which overlapped normal onset of locomotor activity led to large phase advances, rather than modest delays as obtained with nocturnal rodents at that time. There was a very sharp "break point" between timing of stimuli which caused maximum phase delay (200 degrees,

---

[12] The difference is not due simply to the fact that 6-h stimuli were used for the bird, while 10- to 15-min stimuli were used for rodents. Experiments with nocturnal rodents indicate that light stimuli of many hours duration do not usually produce phase shifts of more than about 4 h (cf. Fig. 10.2).

i.e., about 14 h after activity onset) and those which caused maximum phase advance, shortly thereafter; in the rodent data, no such discontinuity is evident.

I have obtained experimental results similar to those of Eskin (1971), when treating another passerine bird, the house finch, with 6-h light stimuli (Enright 1965; Hamner and Enright 1967; Figs. 7.5 and 9.6; and other unpublished results). My data are not as extensive as Eskin's, but because of those similarities, the phase response curve shown in Figure 11.1 will be taken as representative of expectations for a diurnal bird, recognizing that fewer experiments on a smaller number of species are available than for nocturnal rodents (Chap. 10).

11.1 Simulation Results: Several Deficiencies and Their Remedy

A preliminary attempt to obtain comparable results in simulations with the type model, using 6-h decreases in threshold to mimic 6-h light pulses, gave completely unsatisfactory results. A first conspicuous discrepancy was that when a light pulse was administered beginning about 12 h after activity onset, discriminator feedback persisted for many hours after the end of the stimulus. Observations from the house finch demonstrate that this is extremely unrealistic. In many dozens of experiments, with large numbers of individuals, in which single light stimuli were given at this time, locomotor activity of the finch terminated at the end of the stimulus in nearly every case, and did not begin again until many hours later (example in Fig. 9.6; see also Hamner and Enright 1967). A related discrepancy was evident in the simulations of Figure 9.7B and was noted at that time: discriminator feedback persisted long after end of the lighted phase of the stimulus cycles, but this is rarely seen in equivalent experiments with either the house finch or the house sparrow.

Proper simulation of this phenomenon evidently requires more than simply altering the value of a parameter or two in the type model; in seeking a solution to this sort of problem, I have investigated many sets of parameter values which differ from those of the type model, and all were unsatisfactory. Fortunately, an adequate modification of the models can be proposed, which does not require introduction of any fundamentally new concept or mechanism. The supplementary assumption made in Chap. 10 for nocturnal rodents can be reinvoked here: that a sudden lighting-induced increase in discriminator threshold is accompanied by supplementary transient inhibition: an overshoot effect for diurnal birds, comparable with that assumed for nocturnal rodents. This proposal, in the context of a diurnal formulation for responses of threshold to light stimuli, is presented schematically in Figure 11.2.

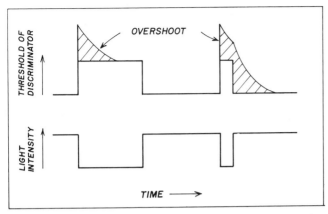

Fig. 11.2. Schematic diagram of the postulated overshoot due to sudden increases in discriminator threshold for a diurnal bird. In simulation of this process, threshold was increased to N, following a step-down in light intensity, and the supplementary effect was permitted to decay exponentially to the level dictated by prevailing intensity, with a half-time of 1 h

Further simulations, after incorporation of this modification into the type model, produced patterns of phase shift which are comparable in many respects with experimental data from the sparrow and the house finch. Four single examples of the simulation results are shown in Figure 11.3, and the resulting phase response curve from a full scanning of the circadian cycle with such stimuli is shown in Figure 11.4. Points of agreement with the experimental data include large-magnitude phase advance resulting from stimuli which begin several hours before activity onset; maximal values of both phase delay and phase advance during the time of inactivity; and what seems to be a very sharp transition from maximum phase delay to maximum advance sometime in the middle of the inactive time. This transition, however, is less interesting than it may initially seem to be. Careful consideration of the individual simulations shown in Figure 11.3 indicates that the apparent discontinuity in Figure 11.4 is a consequence of the somewhat arbitrary decision about whether a given change in subsequent timing of the rhythm is to be regarded as delay or advance. The crossover point in Figure 11.4 follows from interpreting the simulation result of Figure 11.3C as an advance of 12 h rather than a delay of 12 h. Even a phase advance of 4 h can in principle be interpreted as a delay of about 20 h, and therefore the discontinuous crossover point in a phase response curve like that of Figure 11.4 is primarily a matter of convention. (The same problem can arise, in somewhat less acute form, in the experimental data from birds.)

While it is instructive to note such qualitative agreement between these simulations and the experimental data, a major quantitative discrepancy is conspicuous. The range of phase shifts obtained in the simulations (that is,

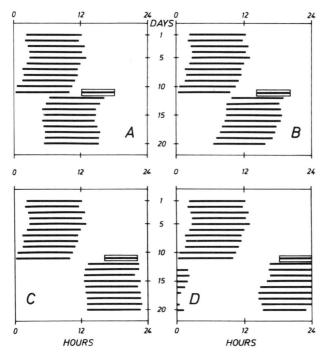

Fig. 11.3 A–D. Simulations illustrating the phase shifts resulting from 6-h threshold-lowering stimuli administered 12, 14, 16 and 18 h after activity onset. Calculations using the type model, and incorporating the overshoot of Figure 11.2, with an identical set of 400 pacers and identical values of z_j up to hour 240. Note that while case **A** appears to be a clear delay in phase, and case **D** to be an advance in phase, cases **B** and **C** demonstrate that these interpretations are rather arbitrary, being dependent upon the interpretation assigned to the forced activity evoked by the stimulus

the range between maximum delay and maximum advance in the curve of Fig. 11.4) was some 20 h, regardless of where one chooses to place the "switchover" from delays to advances. The range obtained in the house-sparrow experiments (Fig. 11.1) was, instead, approximately 12 h (4 h delay, 8 h advance); unpublished data from the house finch suggest an even slightly smaller range for this species.

Trial-and-error attempts to remedy this discrepancy by changing the value of one or another parameter from that of the type model were initially unsuccessful. Eventually, a satisfactory form of the model was arrived at; this required simultaneously reducing the values of $\bar{\delta}$, the Discharge Coefficient; β, the Between-pacer Variability; γ, the Discharge Variability; and ϵ, the Excitation involved in feedback; and – because $\bar{\delta}$ was reduced – increasing the value of $\overline{X}_b$, the Mean Duration of Recovery, so as to keep period of the pacemaker near 24 h. Appendix 11.A deals with the rather complex issue of why these particular changes were chosen, and why they produce the desired effect. Suffice it here to note that such changes lead

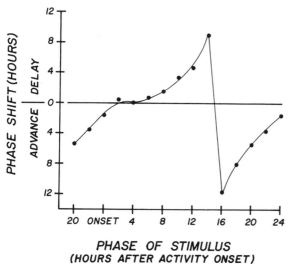

Fig. 11.4. Phase-response curve from 12 simulations like those of Figure 11.3. In this presentation, the results of Figure 11.3B are treated as phase delay, and those of Figure 11.3C are treated as phase advance

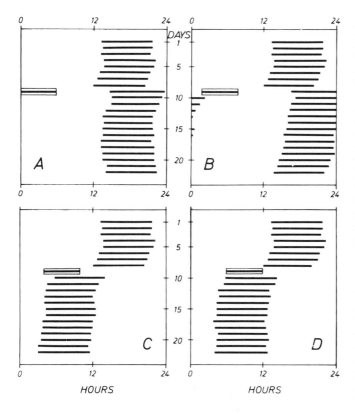

Fig. 11.5 A–D. Simulations showing phase shift resulting from 6-h threshold-lowering stimuli administered 12, 14, 16, and 18 h after activity onset. Calculations based on a model in which $\alpha = 2.5$ h, $\beta = 1$ h, $\gamma = 1/24$, $\delta = 0.4$, $\overline{X}_b = 18$ h, $\epsilon = 5$ h, with 400 pacers in the ensemble; overshoot of threshold (cf. Fig. 11.2) also incorporated. Identical ensemble was used for each case, with identical values of z_j up to the cycle preceding the stimuli. Note the contrast between parts **B** and **C**; a 2-h difference in timing of the stimuli resulted in a 12-h difference in ultimate phase of the rhythm

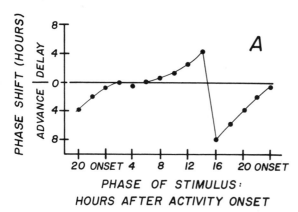

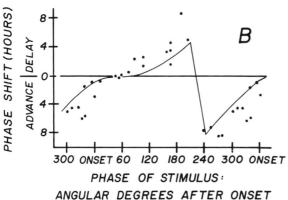

Fig. 11.6. A Phase-response curve based on 12 simulations like those of Figure 11.5, shift being measured on second day after stimulus. B Curve of part A superimposed upon experimentally observed phase shifts obtained in the house sparrow with 6-h stimuli. Data from Eskin (1971)

to a model which satisfactorily reproduces the essential features of the phase response curve of the sparrow. Four single-pulse simulation results are presented in Figure 11.5; the entire phase response curve is shown in Figure 11.6A; and the comparison between that curve and the sparrow data is shown in Figure 11.6B.

Note that Figure 11.6 indicates a 12-h range in the response curve, with maximum delay of about 4 h and maximum advance of about 8 h, in contrast to the 20-h range which was obtained using the type model (Fig. 11.4). A comparison of parts B and C in Figure 11.5 indicates that a 2-h difference in the timing of the stimulus (14 h after onset vs 16 h) led to a 12-h difference in the subsequent timing of the rhythm. This difference provides a somewhat objective rationale for plotting the phase response curve of Figure 11.6 as including a sudden transition from large phase delay to large advance; the same rationale was used in plotting the actual experimental data in Figure 11.1.

The phase delay of the pacemaker rhythm induced by a stimulus administered some 12 or 14 h after onset of feedback is mediated by the unentrained pacers. With basic periods somewhat shorter than that of the ensemble, these pacers — or at least a large fraction of them — will be gradually scanning the latter portion of the inactivity time, and thereby making a contribution to the next onset of feedback, as shown in the upper portion of Figure 11.7. The simulated light stimulus (cycle 5) produces feedback

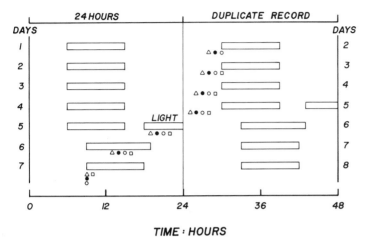

TIME : HOURS

Fig. 11.7. Schematic illustration of the expected onsets of discharge of four unentrained pacers with 23-h period (each pacer shown with a different symbol), during a single 6-h pulse of threshold-lowering light on day 5. The discharge phases of these pacers would, in the absence of the stimulus, each have contributed to onset of feedback on day 6. Because of their acceleration on day 5, they are no longer available to fulfill this role, so onset of feedback will be delayed

at an "unexpected" timing, which serves to accelerate these unentrained pacers. The ensuing onset of feedback is then delayed, because these pacers have been removed from participation in the buildup of discharge preceding onset; attainment of threshold will require a longer waiting time.

When the simulated light stimulus is administered somewhat later in the circadian cycle (parts C and D of Fig. 11.5), the onset of feedback has the capacity of prematurely triggering the discharging of the *entrained* pacers, which would ordinarily have begun their discharge several hours later. This event can serve, then, to produce a massive acceleration of the entire pacemaker ensemble. The switchover in the phase response curve, from delays to advances, depends upon a timing of the stimulus (and a strength of Excitation, ϵ) such that the stimulus-produced feedback is just barely appropriate to affect a large fraction of the entrained pacers, whereas slightly earlier timing cannot.

Note that the phase advance of the pacemaker ensemble resembles the premature triggering of a single relaxation oscillator: onset of the stimulus 2 h before spontaneous onset of feedback accelerates the rhythm by 2 h;

onset of the stimulus 6 h before spontaneous onset advances the rhythm by about 6 h. Figure 7.5 presents experimental data from the house finch in which two such saturation phase advances were induced. A thoughtful examination of this kind of experimental record provides strong intuitive support, I believe, for several features of the general class of models under consideration here, including the emphasis on onset of locomotor activity as a critical phase in the pacemaker cycle, and the assumption that relaxation-oscillator characteristics are an essential aspect of pacemaker behavior.

The simulations of Figure 11.5, parts B and C, suggest the possibility that some sort of modest, single-cycle transients may have accompanied the larger phase shifts of the pacemaker, both advances and delays. These were not, however, a generally reproducible aspect of simulated phase shifts with this formulation of the model, and when they did appear, as in the case of Figure 11.5, they were a minor component of the phase-shifting process, compared with the transients seen in phase advance of nocturnal rodents. There is no indication in the available literature (nor in my experience with the house finch) that diurnal birds consistently show significant transients before completing phase shifts due to 6-h stimuli. (None are evident in Fig. 7.5; nor in Fig. 9.6.) No attempt has been made, therefore, to determine whether the slight transients suggested in Figure 11.5 might be eliminated or emphasized by further modifications in parameter values of the model. Before any such investigation would be fruitful, it would be important to refine the choice of parameters for the model more carefully than was done for the simulations which led to Figures 11.5 and 11.6. Such refinement should probably be based on experimental data from single individual birds. The observations which have been approximately reproduced by the model (Fig. 11.6B) were derived from many different birds, and no doubt include appreciable inter-individual variability, which one should not expect to be adequately represented by a single model with unique assignment of parameter values.

11.2 Phase-response Curves: a Review

The simulation results described in these two chapters demonstrate a striking mirror-image symmetry in the mechanisms by which phase shifts are accomplished in the nocturnal and the diurnal formulations of Coupled Stochastic Systems. Phase delay in the nocturnal formulation (light raises threshold) is accomplished by postponing onset of discharge by pacers of the entrained subset; phase advance, in the diurnal formulation (light lowers threshold) is the converse process, and is accomplished by a strong acceleration of the entrained subset of pacers. Phase advance, in the nocturnal formulation, represents an effect of the stimulus which is mediated by the *un-entrained* subset of pacers, and the same is true of phase delay in the diurnal formulation. With the involvement of unentrained pacers, the processes of interest are significantly more complex. In the nocturnal formulation, cut-off of feedback causes appropriately phased unentrained pacers to postpone their onset of discharge, leading to the initially unexpected result that the pacemaker ensemble will be phase advanced. In the diurnal formulation, phase delay is due to the converse process: appropriately phased members of the unentrained subset are accelerated by the stimulus, and their pre-mature discharge leads to the initially surprising result that the pacemaker ensemble as a whole is delayed. The seeming paradoxes involved in these cases can be satisfactorily resolved by detailed examination of the role of the unentrained pacers in the performance of the ensemble. The results are subtle consequences of the entire coupling process.

The conclusions to which we have come in these evaluations of phase-shifting experiments differ in a conspicuous and important way from most of those in earlier sections. In Chap. 7, it was noted that a whole array of experimental phenomena which have been documented in the performances of nocturnal and diurnal vertebrates under free-running conditions could be satisfactorily reproduced by a broad class of Coupled Stochastic Systems. With the assumption that light raises discriminator threshold for nocturnal animals and lowers threshold for diurnal animals, the type model itself was able to reproduce these phenomena, at least in broad outline. Furthermore, many other models of the general class, created by altering values of the parameters of the type model, can equally well reproduce general phenom-ena such as those summarized by Aschoff's rule. In Chap. 9, we examined the responses of the models to various sorts of light-dark regimes, both square-wave and sinusoidal light cycles. In these cases, as well, a majority of the experimental results can be satisfactorily reproduced by the type model, and also by many other models of the general class, created by alter-ing the values of selected parameters.

These kinds of agreement between model and experimental data provide essential support for the suggestion that Coupled Stochastic Systems can be a useful descriptive vehicle for the data. Since *many* models, however, including the type model, perform satisfactorily in these regards, the use of the models to predict the outcome of novel experiments becomes problema-tic. Which model of the class should be used as the basis for quantitative prediction?

The examination of phase-response curves in Chapters 10 and 11 has served greatly to narrow our choices. Instead of finding that many models of the class can do everything asked of them, we have noted that the type model is conspicuously unsatisfactory for the phase-response curves both of nocturnal rodents and of diurnal birds. The discrepancies were sufficiently serious for a critical supplementary assumption to be required, invoking a short-duration overshoot effect of light stimuli which suddenly raise discriminator threshold. Even this modification has not led to a single broad class of models which might be appropriate for both nocturnal and diurnal animals. A satisfactory fit to experimental data apparently requires that the magnitude of feedback, represented by the parameter ϵ, be *increased* from its type-model value for certain nocturnal rodents, and *decreased* from its type-model value for certain diurnal birds. Furthermore, in both cases, it proved important to reduce the inter-pacer variability in basic period. This was accomplished by reducing β from 3 h to 1 h in both cases, and by reducing γ as well (from $1/8$ to $1/24$) in the diurnal formulation.

A large value for β was initially introduced into the type model to demonstrate that an extremely heterogeneous ensemble of pacers could produce a highly precise output, following the rules of interaction proposed. This kind of extreme heterogeneity, however, has now turned out to be incompatible with experimental phase-response-curve data. The models, therefore, in their context, tell us something very explicit about the necessary underlying phenomena, something which was by no means obvious in the experiments from which the conclusion was derived.

These kinds of restrictions on the class of acceptable models for a given organism, which have resulted from attempts to simulate phase-response-curve data, represent, in addition, important progress toward predictive use of the models. It has now been demonstrated (as should have been expected) that one cannot just arbitrarily select any model of the class, and demand that it give generally applicable quantitative predictions. The values for ϵ, β and $\bar{\delta}$ appropriate to certain species are strongly constrained by existing phase-response data. Because of these constraints, species-specific predictions can be made in much more quantitative detail than would be possible when dealing only with Coupled Stochastic Systems as an entire class.

Appendix 11.A. Constraints on Parameter Values

Careful consideration of Figure 11.7 can contribute materially to the selection of parameter values for a model which should simulate the phase-response-curve data of Figure 11.1. We wish to induce a phase delay of the pacemaker by the process shown in Figure 11.7, using a stimulus which begins, say, 12 h after activity onset. This demands that the stimulus be able to accelerate unentrained pacers such as those illustrated; and simultaneously requires that the terminal portions of the stimulus — at 18 h after onset of activity — not appreciably perturb the *entrained* pacers. If Excitation, ϵ, has its type-model value of 8 h, the feedback operative during the latter portions of the stimulus will be sufficiently strong to engulf most of the entrained pacers. Many of them would ordinarily have begun to discharge spontaneously, only 5 or 6 h after termination of the stimulus, and they would therefore be accelerated by the stimulus. Clearly, therefore, ϵ must be reduced from its type-model value.

Granted that ϵ has been appreciably reduced, so that the stimulus directly affects only unentrained pacers, it is evident that acceleration of those pacers will not lead all of them to begin discharge at the *start* of the 6-h stimulus. Instead, many of those unentrained pacers will be accelerated only to the latter portions of the stimulus, e.g., the pacer shown by open squares in Figure 11.7. Were such a pacer to have a long duration of discharge, it could still effectively contribute to the next attainment of threshold; its being accelerated by the stimulus would not prevent its contribution to the ensuing activity onset, and it would, then, make no contribution to phase delay of the system. Thus we see that unentrained pacers, which are accelerated by the light stimulus, and which have a relatively short duration of discharge, are best suited to induce phase delay of the ensemble. Shortening the duration of discharge by the pacers requires reduction in $\bar{\delta}$, the Discharge Coefficient; but if $\bar{\delta}$ is reduced without a concommitant increase in $\overline{X}_b$ (the Mean Duration of Recovery), then the period of the pacemaker will be reduced and no longer be near 24 h. Hence, if δ is to be reduced, $\overline{X}_b$ must be increased.

This reduction in $\bar{\delta}$ and increase in $\overline{X}_b$ has another salutary effect upon the pacemaker performance. Using the type-model values for these parameters, the minimum value for duration of feedback predicted is some 9 to 10 h (Fig. 6.9). Under experimental conditions which ought to be comparable with such high values of threshold (total darkness for the sparrow, very dim light for the house finch), it is common for birds to show durations of locomotor activity which are appreciably shorter than this type-model duration of feedback. In the sparrow, values of 8 h or slightly less have been observed; values as short as 6 to 7 h are not uncommon in the house finch. Reducing $\bar{\delta}$ to 0.4 and increasing $\overline{X}_b$ to 18 h serves to reduce the minimum attainable duration of pacemaker activity, and thereby provides a better fit to this aspect of the experimental data on the free-running rhythms of birds.

Given the type-model values for β, Between-pacer Variability, and γ, the Discharge Variability, the unentrained pacers have a very broad range of basic periods. Many of them will have periods so much shorter than that of the pacemaker system as a whole, that only a small fraction of their cycles will involve onset of discharge timed appropriately so as to participate in the process sketched in Figure 11.7. Maximal phase delay demands that many of the unentrained pacers – like those of Figure 11.7 – have basic periods only slightly shorter than the period of the ensemble, so that they very gradually scan the inactive portion of the pacemaker cycle. By reducing the values of β and γ, the spread in basic periods of the ensemble can be reduced, with the consequence that pacers of the unentrained subset will have periods only slightly shorter than the free-running period of the pacemaker. In this situation, they will contribute to the summed input which determines activity onset in a larger fraction of their cycles; they would, then, be more readily available to participate in the phase-delay process envisioned here.

The conclusion to be drawn from these considerations — which, in fact, represent my eventual understanding of the pacemaker system, which emerged only after many trial-and-error attempts to simulate a phase-response curve like that of Figure 11.1 — is that the required modifications of the type-model parameter values seemingly include reducing Excitation ϵ, as well as reducing β and γ, the coefficients which determine inter-pacer variability in basic period. Reducing $\bar{\delta}$ also contributes to a large-magnitude phase delay, from appropriately timed stimuli.

Chapter 12
Plasticity in Pacemaker Period: a Dynamic Memory

In experimental studies of the wake-sleep cycle of higher vertebrates under constant conditions, it is common to find that the average free-running period of an animal's rhythm, under a given set of conditions, varies to a small extent from one test of many cycles to the next. Recordings of the performance of a single animal over many months often show gradual trends toward shortening or lengthening of period; and it is not uncommon to observe sudden stepwise changes in period, sometimes caused by a known perturbation of the animal, occasionally due to unknown causes ("apparently spontaneous"). Such differences demonstrate, by definition, that the period of the circadian pacemaker is not uniquely determined by concurrent environmental conditions; in some way or another, the animal's history can affect its circadian performance.

Pittendrigh (1961) was apparently the first to note that some of this long-term variability in period of a circadian rhythm is predictable, and is apparently caused by preceding exposure of the animal to certain sorts of light treatment. For such systematically inducible alterations in circadian period, he proposed the term after-effects. Pittendrigh and Daan (1976a) have recently reviewed the literature on after-effects and presented much new data on the topic, predominantly but not exclusively from nocturnal rodents. The phenomena reported include after-effects due to period (i.e., cycle length) of an entraining light regime; due to photoperiod of a light regime (hours of light per day); due to single phase-shifting stimuli; and due to treatment with constant light.

One of the ways in which past events can modify pacemaker performance for several days has already been encountered in the treatment of transient cycles associated with phase advance of nocturnal rodents (Chap. 10). The sorts of history dependence in circadian rhythms of interest in this chapter, however, go far beyond the short-term changes involved in these transients, and represent alterations in pacemaker performance which persist for weeks and even months.

A question of considerable importance, in terms of physiological mechanism, is whether there is a genuine qualitative difference between after-effects and the transients associated with phase shifting. Are after-effects no more than very-long-term transients? In introducing the concept of after-effects, Pittendrigh (1961) interpreted them as induced changes in period at *steady state*. "The system can be pushed within this range (of realizable period values) to any one of, presumably, many frequencies where it is stable, at least for a while; its state perpetuates itself." (ibid., p 167). In the most recent treatment of the topic, however, the emphasis on steady state is missing. ". . . after-effects are long-lasting, slowly decaying, changes in pacemaker period incurred by (*sic*) prior experience." (Pittendrigh and Daan 1976a, p 234) The basis for this latter definition is embodied in the finding that while the initial after-effect of a given pretreatment may be relatively large, the induced change in period gradually becomes smaller with the passage of time. Such behavior suggests the possibility that, given sufficient time, the pacemakers of the animals might eventually converge upon some unique value of free-running period, which would represent *the* steady-state behavior of the system under the prevailing constant conditions. Were that to be so, then after-effects would represent long-duration but transitory phenomena, as the pacemaker gradually moves toward the steady state.

Although this kind of decay shows beyond doubt that the major component of most after-effects is indeed transitory, small residual effects have sometimes been noted which may indeed represent true differences in steady-state behavior. That possibility will be considered in detail later in this chapter; for the moment, however, we will focus attention on the larger-magnitude after-effects which very gradually decay.

In a broader context, persistent and even permanent changes in the behavior of an animal are, of course, neither novel nor surprising; learning and conditioning are familiar examples of history dependence, as are behavioral changes due to maturity and to aging. Such modifications of responsiveness are in most cases presumed to be associated with alterations in physical substrate: modification of synapses, growth of organs, changes in hormone levels or the like.

Based on that background, it would be a simple matter to propose an ad hoc elaboration of Coupled Stochastic Systems, so as to force them into conformity with various reported after-effects in free-running period. The obvious procedure would be to postulate an inducible change in pacemaker anatomy: some alteration, due to experience, in the value of one or more of the parameters which specify the intrinsic properties and determine the behavior of the pacemaker system. For after-effects which deteriorate over several weeks, one could introduce a "forgetting" effect, in the form of a decay process with a very long time constant. It turns out, however, that this sort of elaboration of the models is unnecessary. Certain models of the class can account for the observed plasticity of circadian period in considerable detail as a natural consequence of their normal functioning, without postulating that experience induces changes in pacemaker parameters or making any additional assumptions whatever. Circadian after-effects thus constitute a kind of long-term memory, which can be adequately accounted for without invoking modifications in physical substrate for the storage of information.

12.1 After-effects of Initial Phase Conditions

All computer simulations described to this point were initiated with the entire population of pacers placed at an identical phase; each simulation began with all pacers at the start of their recovery phases. A thorough examination of the behavior of the type model demonstrated to my satisfaction that this calculational procedure had no serious drawbacks. As was noted in the discussion of Figure 5.3, for example, the behavior of the system in circadian cycle 1 differed very slightly from cycle 2, but cycle 15 appears to be equivalent to cycle 2 in all essential details. The possibility that the set of phases imposed on the pacers as starting conditions might alter system performance is a disturbing possibility which was frequently considered and reexamined during the computer studies described previously. In all cases, it appeared that transient behavior resulting from these starting conditions had satisfactorily dissipated within two to three circadian cycles at the most. In further exploration of this ground for concern, other simulations demonstrated that the imposition of radically different starting conditions, with, for example, a very broad distribution of initial phases, did not detectably alter system performance.

This is not to say that the type model and its variants consistently behaved as simple relaxation oscillators, with no history dependence. In most free-running simulations, the coefficient of serial correlation between period length in successive cycles was positive; and these correlations were "statistically significant" far more often than could be attributed to chance alone. Two extreme examples of the kinds of phenomenon which can underlie these correlations in the simulations are illustrated in Figure 12.1. In such cases, the period of the pacemaker seems to evolve, in a rather erratic fashion, presumably due to occasional unusual "events" resulting from the operation of stochastic processes. None of these striking phenomena was, however, consistent or predictable in replicate simulations, using different sequences of stochastic variables (values of z_j). The type model apparently shows only very modest short-term correlations in its cycle-to-cycle behavior. In view of the magnitude of stochastic variability incorporated into the models, it is difficult to imagine how the situation could be otherwise; any peculiarities in system performance which might arise due to the initial phase relationships among pacers ought to be degraded very rapidly.

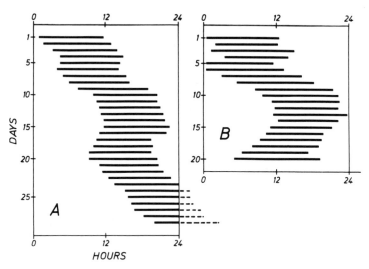

Fig. 12.1 A and B. Two simulations with the type model, in which there was very strong serial correlation in period of the system during successive cycles. **A** an "oscillation" in free-running period; **B** an apparent shortening of free-running period after about day 14

Very late in my studies of Coupled Stochastic Systems, however, I discovered that with the appropriate choice of parameter values, pacemaker period *can* be influenced by the initial phasing imposed upon the pacers. Data showing such results are presented in Figure 12.2: duplicate simulations for each of four pacemaker ensembles. The values assigned to the parameters of the model were in all cases those given in the figure legend, but each of the four ensembles was created by use of a different array of Gaussian variates ($z_{i,\beta}$ and $z_{i,\gamma}$) to specify the properties of the individual pacers, leading to modest differences among the four pacemakers in details of anatomy. The duplicate simulations (two sequences of points for each ensemble, grouped together under brackets A, B, C, and D) were calculated by using independent sequences of stochastic variables (values of z_j). Note that in each simulation in the figure, the free-running period of the system was somewhat shorter during the initial cycles than the value finally attained; there was a *very* gradual lengthening of free-running period.[13] Each of these

---

13 Small but clear differences are evident between the performances of the several pacemaker ensembles of Figure 12.2, differences which were repeatable in the duplicate simulations. For example, the final portions of the two simulations using ensemble A (sequences on left side of figure) indicate that this pacemaker tended to evolve a slightly shorter free-running period than ensemble D (sequences on right side of figure). Note also that ensemble D tended to increase in period somewhat more than either ensemble A or B. The differences between average period in the first and last

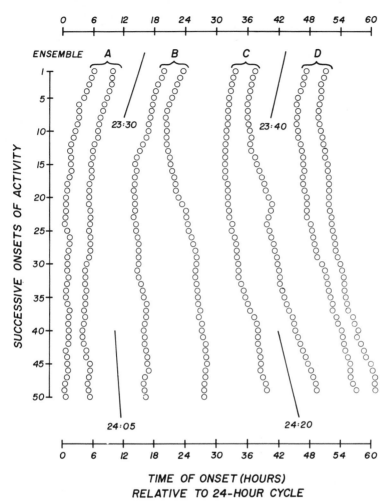

Fig. 12.2. Sequences of activity onsets obtained from eight free-running simulations, all with models in which $\alpha = 1$ h, $\beta = 1$ h, $\gamma = 1/80$, $\bar{\delta} = 0.2$, $\epsilon = 3$ h, $\overline{X}_b = 20.5$ h, $N = 200$ and $\theta = N/4$; all pacers assigned an identical starting phase for initiation of the calculations. Four pacemaker ensembles, designated A, B, C, and D, were used, as determined by four arrays of values for $z_{i,\beta}$ and $z_{i,\gamma}$. Duplicate simulations (i.e., different sequences of values for z_j) were performed with each ensemble. See text for further discussion

"weeks" of simulation were 33 and 37 min for ensemble D; 23 and 26 min for ensemble A; and 22 and 26 min for ensemble B. Such differences apparently reflect chance-dictated details of pacemaker anatomy, arising from the fact that finite samples of the Gaussian distribution were used to specify properties of the individual pacers of a pacemaker. The differences between duplicate simulations in Figure 12.2 reflect the intrinsic, inescapable unpredictability of a Coupled Stochastic System; the differences in results between ensembles, however, can be interpreted as reflecting the equivalent of inter-animal variability as well.

simulations was initiated with all pacers in the ensemble at an identical
phase. The gradual lengthening of pacemaker period during the simulations
was associated with the gradual deterioration of that imposed close phase
synchrony.

After-effects of the sort of interest in this chapter arise when particular
kinds of light treatment lead to changes in period of the pacemaker. The
results of Figure 12.2 do not directly demonstrate that kind of phenomenon,
since there were no differences in the starting conditions of the various simulations. Instead, these calculations only demonstrate that changes in the
phase relationships among the pacers, which gradually evolve during long
simulation, can alter period of the pacemaker. Nevertheless, with the demonstration (see below) that given kinds of pretreatment can induce comparable differences in phasing of the pacers, such results have a direct bearing on after-effects.

Before we proceed in that direction, however, it is important to understand the origin of the long-term trends in pacemaker period shown in Figure 12.2. A first step toward this objective is to establish the range of values
of parameters which lead to those results, which were never obtained with
the type model. On the basis of extensive simulations, it appears that results like those of Figure 12.2 depend upon fulfillment of all the following
requirements: the range of basic periods in the ensemble must be relatively
small, meaning that β and γ should be smaller than in the type model; and
α, ϵ and $\bar{\delta}$ also must have values appreciably smaller than those of the type
model. With these requirements in mind, let us consider the idealized ex-
pected behavior (i.e., neglecting stochastic variation) for unentrained pacers,
such as the one illustrated in Figure 12.3. In a simulation which begins with
all pacers initially in phase with each other, unentrained pacers with period
near that of the ensemble will gradually drift forward in phase, relative to
discriminator activity. Eventually, their discharge will no longer overlap
onset of feedback, at which point (cycle 7 in Fig. 12.3) they cease to contribute to attainment of threshold. This would, then, serve slightly to length-
en the period of the ensemble, in the same manner seen in previous chapters: a longer waiting time will be required for summed discharge of the
remaining pacers to exceed threshold. If ϵ is small, an unentrained pacer
can continue for many cycles thereafter to produce its discharge during the
inactive time (cycles 7 to 18 in Fig. 12.3); during this entire interval, the
pacer is "lost" in terms of system performance. The unentrained pacer
shown in this example would on the average contribute to attainment of
threshold in only about 30% of its cycles.

With all the pacers initially in phase with each other, there will also be
an element of self-catalysis in this process; when the period of the ensemble
is slightly lengthened by the loss of some of the unentrained pacers, then
other pacers, which originally were just barely entrained (with basic periods

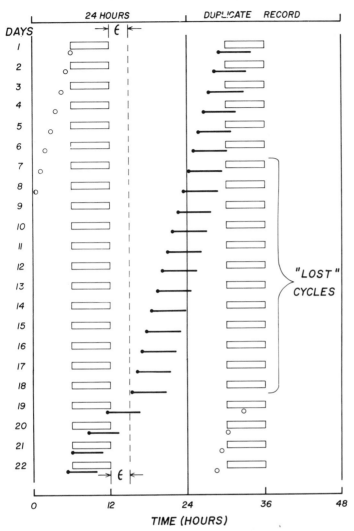

Fig. 12.3. Schematic illustration of the expected sequence of discharges of an unentrain-ed pacer with a period of 23.2 h, neglecting stochastic variation. Discriminator feedback activity indicated by *open rectangles;* discharge of the pacer indicated by *solid horizontal lines.* Note that between days 7 and 18 (lost cycles) discharge of this pacer would have no effect on discriminator activity

very slightly longer than that of the ensemble) will no longer be entrainable, and they, too, will begin gradually to scan the inactive portion of ensemble behavior, eventually also disappearing into the inactive time. Since these latter pacers would have basic periods extremely close to that of the en-semble, they would all require many cycles to complete the process shown in Figure 12.3.

Careful consideration of this idealized example indicates why the gradual increase in pacemaker period depends upon small values for β and γ, as well as $\bar{\delta}$, ϵ and α. The very slow scanning of discriminator cycles illustrated in this diagram depends on the fact that the pacer has a period only slightly shorter than that of the ensemble. For such behavior to have a significant effect upon the ensemble, there must be many unentrained pacers with such values of basic period. In other words, the total range of basic period values must be relatively narrow; small values of β and γ are necessary for fulfillment of that requirement. If ϵ were to be large, the unentrained pacers would be accelerated to the end of the preceding block of feedback far sooner. For example, with ϵ of 10 h instead of 3 h, the "jump forward", which occurred in cycle 19 in Figure 12.3, would have occurred in cycle 10 instead. The pacer would, then, on the average, be expected to contribute to attainment of threshold in 50% of its cycles, instead of only 30%. In essence, then, a large value of ϵ reduces the duration of the interval during which an unentrained pacer remains "invisible" in system performance. If $\bar{\delta}$ were to be appreciably increased, the duration of discriminator activity (open rectangles in Fig. 12.3) would be lengthened, and the duration of discharge by the unentrained pacer (solid horizontal lines) would be lengthened as well; both of these effects would lead the unentrained pacer to contribute to onset of feedback in a larger fraction of its cycles. The requirement that stochastic variability also be relatively small, in order to obtain these long-term trends in period, depends upon more subtle considerations. One might suspect that if α were large, an equivalent average result could be obtained simply by increasing the number of pacers in the ensemble, but that is not entirely correct. Large stochastic variation would mean that a pacer like that shown in Figure 12.3 would complete its scan in applicably fewer cycles because stochastic variation alone would frequently lead to earlier crossing over the "boundary" set by ϵ. The pacer might thereby reach the phase shown in cycle 19 as early as, say, cycle 14. With a large value for α, therefore, the unentrained pacers would have a larger fraction of their cycles phased such that their discharge contributes to onset of feedback. In effect, a large value for α has the same influence on the process shown in Figure 12.3 as a modest increase in ϵ.

While the long-term trends in average period illustrated in Figure 12.2 can thus be understood in terms of the expected behavior of the unentrained pacers, the phenomenon illustrated there may initially seem to be irrelevant to actual experimental situations. Those simulation results are a consequence of the artificial and unrealistic "starting conditions", in which all pacers were initially forced into phase synchrony. A comparably extreme distribution of phases among pacers need not, however, be arranged so artificially; trends in this direction (or away from it) are to be expected from

simulation of a variety of simple experimental manipulations. Such cases
are considered in the next sections.

12.2 After-effects of Entrainment

The most familiar, consistent and well documented of the general class
of phenomena known as after-effects are those resulting from experimental
entrainment by light-dark regimes which have a period deviating by an hour
or more from the animal's free-running period. The animal's rhythm during
subsequent free run in constant conditions tends to have a period with a
small bias in the direction of the preceding entrainment regime, with short-
period entrainment leading to a rhythm with period which is shorter than
that following long-period entrainment (Pittendrigh and Daan 1976a and
references cited there). The differences in free-running period inducible in
such experiments can initially be as large as 40 to 60 min, but there is a
clear tendency for this large initial difference to decay gradually over a matter of weeks. How might such a phenomenon arise in a Coupled Stochastic
System?

Consider what happens when the free-running period of the model pacemaker is shortened by an entrainment regime. Any pacer with basic period
longer than that of the ensemble will ordinarily belong to the entrained
subset; entrainment of the pacemaker by a light regime which shortens its
period will increase the fraction of pacers in the ensemble which are entrained, and will leave fewer pacers unentrained. As was shown in Chap. 5
(Fig. 5.6), entrained pacers will ordinarily have discharge phases which begin either just shortly before or shortly after onset of feedback. Hence,
entrainment of the pacemaker to a short value of period can be expected
to produce a distribution of phases of most of the pacers which roughly
resembles the starting distribution in the simulations illustrated in Figure
12.2. If the pacemaker has parameter values similar to those involved in
Figure 12.2, one ought therefore to expect a long-duration trend in period
resembling those illustrated in Figure 12.2, when the pacemaker free-runs
following such short-period entrainment. On the other hand, entrainment
by a stimulus regime which lengthens period of the system will entrain a
smaller fraction of the pacers, and will be associated with considerable dis-
persion in the phases of the unentrained pacers, a situation resembling that
finally attained, after prolonged free run, in Figure 12.2. When the pacemaker is thereafter permitted to free-run under standard conditions, it
should proceed relatively rapidly an equilibrium distribution of phases by
the unentrained pacers.

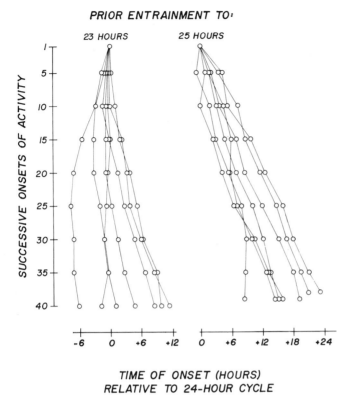

Fig. 12.4. Plots of every fifth successive onset of discriminator activity from simulations with seven pacemaker ensembles, each with parameter values as given in the legend to Figure 12.2. All pacers of an ensemble were assigned identical phasing at the start of the simulation, and for the first 10 days an entrainment regime was imposed upon threshold, resulting in a pacemaker period of either 23 h of 25 h. Following that entrainment, the simulations were continued for another 40 days of free run, to produce these data. Each ensemble was used for both 23- and 25-h entrainment

Confirmation of these expectations is illustrated in the results shown in Figure 12.4. Data from seven different pacemakers, including the four of Figure 12.2, are shown there, each pacemaker determined by a different array of values for $z_{i,\beta}$ and $z_{i,\gamma}$. Each pacemaker was subjected, for the first 10 days of a simulation, to a stimulus regime (square wave of threshold) with a period of 23 h, and then permitted to free-run (results illustrated in the left half of the figure). Then the pacemaker was returned to the same starting conditions (all pacers in phase synchrony), entrained for 10 days to a stimulus regime with period of 25 h, and then permitted to free-run (results illustrated in the right half of the figure). Entrainment of the pacemaker to a short-period stimulus regime resulted in a conspicuously shorter average period, during the initial stages of subsequent free-run, than did entrainment to a long-period regime. These effects persisted for at least 3 "weeks" of simulated free-run, with gradually decreasing magnitude of the residual after-effect. These data therefore demonstrate that Coupled Stochastic Systems can show long-persistent after-effects due to an entrainment regime which are comparable with those seen in the experimental data. Whether the pacemakers in these calculations had achieved steady state after 6 weeks of simulation, and if so, whether there were residual steady-state differences in period, are issues which will be considered later.

The initial differences in pacemaker period shown in Figure 12.4, due to short-period and long-period entrainment, are on the order of 30 to 40 min. Those values are in reasonably good agreement with corresponding experimental data, although intergroup differences of more than an hour were observed of *Mus musculus* following entrainment to 20-h and 28-h light cycles (Pittendrigh and Daan, 1976a, Fig. 8). The rate at which the after-effects of entrainment decayed in Figure 12.4 appears to be slightly more rapid than in comparable experimental data. Most of the effect shown in these simulations had dissipated within about 4 weeks of free run; somewhat longer duration of decay has occasionally been documented in the animal experiments. On the basis of the interpretation of Figure 12.3, I believe that longer-enduring after-effects than those of Figure 12.4 could be induced in computer simulation by further reduction in the parameters α and β, the Stochastic Coefficient and the Between-pacer Variability, respectively. Simulations to confirm this intuition, however, have not yet been undertaken.

12.3 After-effects of Constant Light Intensity

When nocturnal rodents are exposed to constant light, the free-running periods of their rhythms are consistently longer than is typically seen in constant darkness (Aschoff's rule; see Chap. 7). Pittendrigh and Daan (1976a) have demonstrated that treatment of the deermouse *Peromyscus leucopus* with constant light consistently leads also to after-effects in the subsequent free-running period; the rhythms of the animals have an unusually long period for at least 30 days following such pretreatment. Additional data on this kind of result are shown in the measurements on pocket mice *Perognathus longimembris* shown in part E of Figure 7.2; following 3 weeks in constant light, the free-running periods of the animals' rhythms in darkness were lengthened in eight of the nine animals tested, with a median increase of about 18 min.

This effect can also be simulated by Coupled Stochastic Systems; the process by which it comes about in the model is quite similar to that involved in after-effects due to entrainment. The influence of constant light on noc-

turnal animals is simulated by a high value of threshold, and constant dark
by a low value of threshold. A high value of threshold produces a larger
value of free-running period than that resulting from a low value of threshold. These differences in free-running period of the pacemaker, induced
by different constant conditions, have effects upon phasing of the pacers
which are roughly comparable with the effects of simulated entrainment
by long-period or short-period threshold regimes. In the short-period rhythm,
most of the pacers are fully entrained, with onset of discharge near onset
of feedback; in the long-period rhythm there are many unentrained pacers,
which achieve a broad distribution of phases. Hence, pretreatment of the
model pacemaker with a high value of threshold ought to lead to greater
values of period during subsequent free run than pretreatment with a low
value of threshold.

Simulations showing this kind of phenomenon are illustrated in Figure
12.5. Results from seven different pacemakers are shown there, as in Figure 12.4, with each pacemaker having been pretreated once with a low value
of threshold and once with a high value. The after-effects of these stimulus
conditions upon free-running period persisted for at least 3 weeks following treatment.

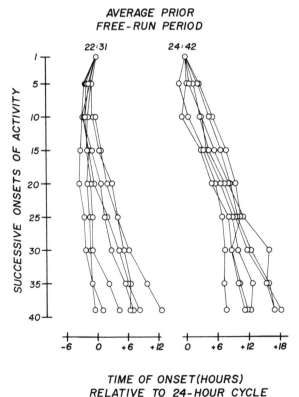

AVERAGE PRIOR
FREE-RUN PERIOD

TIME OF ONSET(HOURS)
RELATIVE TO 24-HOUR CYCLE

Fig. 12.5. Plots of every fifth
successive onset of discriminator activity from simulations
with seven pacemaker models,
all with parameter values as
given in the legend to Figure
12.2. For the first 10 days of
the simulations a steady value
of either 0.15N or 0.4N was
imposed as threshold, resulting
in a short-period (*left data
series*) or a long-period (*right
data series*) free run. The values,
22:31 and 24:42, represent
the seven-pacer average values
of the free-running period attained during the 10-day pretreatments. Following that
pretreatment, threshold was
returned to 0.25N and the
simulations were continued
for another 40 days of free
run, to produce these data.
Each ensemble was used for
both low-threshold and high-
threshold pretreatment

12.4 After-effects of Phase Shifting and of Photoperiod

In their extensive consideration of after-effects seen in the data for nocturnal rodents, Pittendrigh and Daan (1976a) have demonstrated the occurrence of very small but predictable changes in free-running period of a
rhythm induced by single phase-shifting stimuli; and to changes in period
induced by photoperiod (i.e., the number of hours of light in a 24 h cycle).
Simulations with Coupled Stochastic Systems have not yet conclusively
demonstrated that the models show comparable after-effects to these sorts
of stimuli, but Appendix 12.A focusses upon such phenomena. It is shown
there that while simulations much more extensive than those undertaken
here would be required to demonstrate that the models *in fact* can reproduce these kinds of after-effects, there is clear theoretical basis for expecting that Coupled Stochastic Systems *ought* to show small after-effects due
to phase-shifting stimuli, the direction and relative magnitudes of which
agree with the experimental data. Furthermore, there is a theoretical basis
for expecting that photoperiod can induce after-effects in the models,
whenever the number of hours of light per day is insufficient to accommodate the duration of discriminator activity which arises under free-running
conditions.

12.5 Permanent vs Transitory Differences in Period

An evaluation of the question of whether an oscillatory system is or is
not showing steady-state behavior is fraught with both theoretical and practical difficulties. Consider as a first example the activity recording illustrated
in Figure 12.6, in which the rhythm of a house finch was phase-advanced
by a 2-h light stimulus. A replot of the times of activity onset (lower portion of the figure) indicates that following two or three cycles of transient
behavior, the period of the rhythm was shortened by about 12 min. Was
this change in period a transitory effect or did it reflect a new and different
steady-state behavior? An examination of the "before" and "after" portions of the original recording provides no clear reason to suspect that either
portion of the record shows anything other than a steady state. Furthermore, in the replotted data on times of activity onset, there is no indication in the pretreatment data that the rhythm was tending toward an equilibrium value of period shorter than observed, nor any indication in the
posttreatment data that the rhythm was tending toward a longer period.
In fact, however, no rigorous conclusion about steady state is admissible;
all we know is that if some unique steady-state value of period for this bird's
rhythm existed, the tendency to converge on that value was too gradual

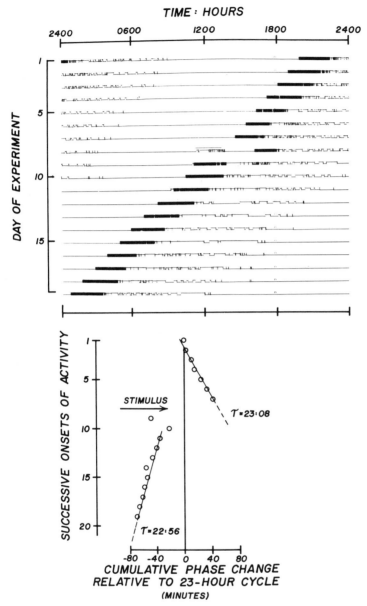

Fig. 12.6. *Upper portion*: activity record of a house finch, subjected to a single 2-h pulse of bright light (11:30–13:30 on day 8) during free run. The shift of phase of the rhythm is conspicuous. *Lower portion*: activity onsets from this actogram, replotted relative to 23-h cycle, to show change in period of the rhythm following stimulus. The *superimposed diagonal lines* indicate that period shortened by about 12 min. See text for further discussion

to be detected in the data available. This particular record is decidedly too short for detection of a trend in period such as shown by some nocturnal rodents, in which the average rate of change of period may be only a fraction of a minute per day.

Consider, then, the record of Figure 7.3; there a discrete and sudden change of about 30 min in the period of the bird's rhythm is shown, and there is no sign in the data of convergence or monotonic trend in the period values during the month's recording before or the month's recording after the sudden change. The data are consistent with the idea of alternative steady states in the bird's pacemaker. Nevertheless, it is also conceivable that the observed 30-min change in period was transitory; that, given sufficient time, a *very* gradual process of convergence to a unique steady-state value would have occurred.

Even stronger evidence for induced change in steady-state period has been provided by Pittendrigh and Daan (1976a). When two groups of mice were entrained to different light cycles, one to a 20-h period, the other to a 28-h period, a difference in average free-running period of the two groups was still detectable (p = 0.04, one-sided rank-sum test) 150 to 200 days after entrainment (ibid., Fig. 8). Furthermore, the intergroup difference in period was about 25 min, measured 50 to 100 days after entrainment; about 19 min, measured 100 to 150 days after treatment; and about 22 min, measured 150 to 200 days after treatment. That is, there was no evidence of trend for the two groups to converge, during the last 4 months of free run. If these residual differences in period do not reflect true steady-state effects, it must at least be conceded that the rate at which the after-effects decay can be extremely small. Other published individual actograms (e.g., ibid., Fig. 12) also demonstrate that a given kind of pretreatment can induce changes in the circadian period of certain animals which appear to be *extremely* stable: so stable, in fact, that one can legitimately doubt whether either of the two observed values of period would spontaneously converge on the other, even if the recording had been continued for the lifetime of the animal.

Consideration of these examples shows nevertheless that a completely rigorous demonstration of alternative steady-state values of period for a circadian pacemaker, based on experimental data, is difficult to imagine. The most that can be said with confidence is that a good deal of evidence is consistent with the idea of different steady states. It is therefore of interest to know whether Coupled Stochastic Systems also show indications of stable or metastable alternative values of period.

A first line of evidence, suggesting that the models may also resemble the animals in this property, is contained in Figure 12.7, based on simulations with two of the four pacemaker ensembles used in Figure 12.2. The final

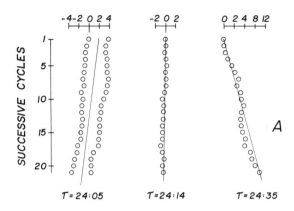

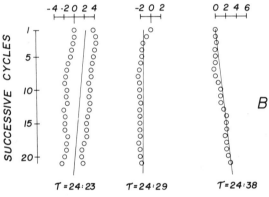

Fig. 12.7 A and B. Successive onsets of discriminator activity from days 30 to 50, showing control runs, and the consequences of prior entrainment. Results are for two of the pacemaker ensembles used for Figure 12.2: ensemble B (upper portion of figure, labeled **A**) and ensemble D (lower portion of figure labeled **B**). *Left columns of points* represent duplicate, 50-cycle free-run simulations, starting with all pacers initially in phase synchrony; these points are directly replotted from Figure 12.2. *Middle columns:* the first 10 days of simulation were under 23-h entrainment; *right columns:* the first 10 days of simulation were under 25-h entrainment. Less detailed results from these latter simulations are illustrated in Figure 12.4. *Lines through the points* represent estimates of average period, with value given in hours and minutes

3 weeks of the duplicate free-run simulations, starting with all pacers in phase, are shown in the left portions of the illustration; the middle and right-hand columns of points resulted from the final 3 weeks of calcula-

tions, following prior synchronization of the same pacemakers to 23-h, and to 25-h entrainment regimes.[14]

There are two issues of interest in these simulation results: (1) differences were present in average period after 3 to 4 weeks of free run, differences which are appreciably greater between pretreatments than within treatment (i.e., within duplicate simulations); and (2) there is no evidence for further long-term trend in period in the simulations, such as characterizes the initial stages (illustrated in Figs. 12.2 and 12.4). That is, the simulations give no reason to suspect that each pacemaker would eventually converge upon some unique value of period, regardless of pretreatment. In spite of weak oscillatory trends in period value, the residual, treatment-induced differences in period *seem* to be stable; the simulation results *appear* to represent steady-state behavior.

As with the data from animal experiments, however, we are confronted here with the difficulty of deciding what criteria would be adequate for rigorous distinction between true alternative steady states, and transitory states which would eventually converge upon a unique, truly steady state. Ultimately, a distinction of that sort must take into consideration the capacity of the system to recover following perturbation, as well as the magnitude of perturbation to which the system is subject on a continual basis. This issue is illustrated by simulations of another sort, described below.

12.6 Bistability of a Pacemaker

The initial idea behind the calculations illustrated in Figure 12.8 was that if an ensemble of circadian pacers in a Coupled Stochastic System were to be initially segregated into two groups, 180° out of phase with each other, a stable circadian pattern might result with two "peaks of activity" per circadian cycle, as seen in the tidal rhythms of certain marine crustaceans. I am now convinced that a model of this sort is inadequate to deal

---

14 Qualitatively similar results were also obtained with the other two pacemaker ensembles of Figure 12.2. With ensemble A, the average period during the last 3 weeks, following initial entrainment to 25 h, was 26 min longer than the corresponding average value of the two control simulations shown in Figure 12.2, and was 23 min longer than the longer of those two control values. With ensemble C of Figure 12.2, the final 3 weeks following 25-h entrainment had an average period 9 min longer than the average of the two control simulations, and 5 min longer than the longer of the two control values. Furthermore, none of those simulations showed any clear monotonic trend in period, during the final 3 weeks of simulation.

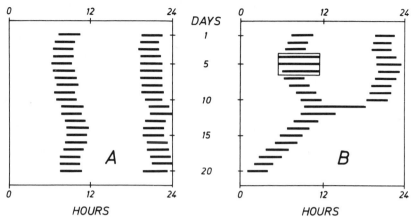

Fig. 12.8. A Free-running simulation with a model in which $\alpha = 1.25$ h, $\beta = 1$ h, $\gamma = 1/40$, $\bar{\delta} = 0.1$, $\overline{X}_b = 26$ h, $\epsilon = 5$ h; 400 pacers in the ensemble, with $\theta = N/10$. The pacers were divided into two groups, and the groups assigned initial phases 12 h different. Note that the bimodal pattern persisted with a free-running period of about 24 h. **B** Simulation with identical ensemble of pacers and identical values of z_j up to hour 80, but with three threshold-lowering stimuli (*enclosed rectangle*) administered on days 4, 5, and 6. Note that following several transients the bimodal pattern coalesced, and that thereafter the free-running period was about 23 h

with the long-term complexity of pattern seen in a tidal rhythm (Enright 1972), unless extensive elaboration of the assumptions be undertaken, which will not be proposed here. Nevertheless, the results of Figure 12.8 illustrate an important property of the models, which was initially unexpected.

In the first simulation (Part A), a bimodal circadian pattern of discriminator activity did indeed emerge, with a period of approximately 24 h. In the second simulation shown in part B (using identical pacers and identical starting conditions), three 6-h perturbations (lowered threshold) were imposed. Following several cycles of complex transient behavior, the pacemaker settled into a unimodal pattern, with a much shorter period — about 23 h.

Other simulations like those in Part A have persuaded me that the bimodal, long-period pattern shown there does indeed represent a stable state of the pacemaker, one which is able to resist all those small perturbations occasioned by the stochastic component of the model, z_j. Following *large* perturbations like those shown in part B, however, the system has a high probability of collapsing into a unimodal state, like that of the simulation shown. In a general sense, that short-period, unimodal state of the system is *more* stable than the bimodal state. As far as I have been able to determine, large perturbations like those used in Part B cannot return the system to its bimodal state, once the unimodal state has been attained. Clearly, therefore, the definitional distinction between metastable and stable states of an oscillator must include consideration of the noise to which the system is subjected.

The simulations of Figure 12.8 provide persuasive evidence that at least under some circumstances, Coupled Stochastic Systems have alternative steady states, characterized by different values of period. The period of the system is not uniquely determined by its anatomy, as specified by its parameter values (including values of $z_{i,\beta}$ and $z_{i,\gamma}$), but can be affected by its history. As is evident in Figure 12.8 — and as follows also from other simulations of this chapter — the phasing of the pacers in the ensemble relative to each other can strongly influence pacemaker period. We have seen several examples in preceding chapters of a reciprocal phenomenon: when period of the pacemaker is altered, for example, by entrainment, the phase relationships among the pacers are altered (e.g., Figs. 9.8 and 9.9). These two relationships — pacer phasing affects free-running period of the system, and period of the system in turn affects pacer phasing — provide the foundation for a dynamic kind of memory. "Information" can be stored for long intervals in the phase relationships among the pacers. Given an absence of large perturbations (Fig. 12.8A), that system memory can persist indefinitely. It is a kind of memory, a history dependence, which does not arise from any alteration in physical structure. There is no localized "engram"; instead, the memory is "distributed"; it resides in the temporal relationships, the very existence of which depends upon the whole ensemble.

At first glance, this phenomenon may seem to constitute only an elaboration of the elementary idea that two highly precise oscillators can show regular "beats" at long-term intervals. The simulations described in this chapter, however, are not based on highly precise systems; they included major stochastic variability in performance of the elements. The intrinsic precision of an individual pacer was no better than 1 part in 25. Even in the face of this high noise level, the system memory — at least in the simulation of Figure 12.8, and possibly elsewhere — shows no sign of progressive degradation.

Another interesting aspect of the kind of memory involved in Coupled Stochastic Systems is that the pacemaker can be "shaped" by its experience. As shown in Figures 12.4 and 12.5, the ensemble tends toward *repetition* of that behavior which has been evoked by recent events. In a very loose sense, one might say that the pacemaker has some capacity to "learn", to show environmentally induced modification of its performance which is potentially adaptive.

The simulations of Figure 12.8, and the bistability indicated there, might easily be dismissed as peculiar consequences of the extremely artificial starting conditions. Nevertheless, the simulation results shown in Figure

12.7, as discussed above, raise the distinct possibility that Coupled Stochastic Systems may have stable — or at least metastable — alternative values of period, even in the unimodal state, without imposing the extreme conditions which led to Figure 12.8. Further consideration of this important possibility from a theoretical viewpoint is contained in Appendix 12.B. It is shown there that changes in the steady-state value of period ought to be expected, following large perturbations of a Coupled Stochastic System, provided that the range of basic periods of the pacers in the ensemble is sufficiently small.

Appendix 12.A. After-effects of Phase Shifting and of Photoperiod

12.A.1 Single Shifts of Phase

On the basis of a large number of phase-shifting experiments with four species of nocturnal rodents, Pittendrigh and Daan (1976a) have demonstrated the occurrence of very small, systematic changes in free-running period induced by single light pulses. Light stimuli which led to large phase advances (more than 40 min) tended to be associated with very slight shortening of subsequent free-running period, and large phase delays with slight lengthening of period. These after-effects were much smaller than those due to entrainment or to constant light: an average change in period of 6 min in the case of phase advance of *Mus musculus,* and much less in all other cases. In each of the four species examined, the average change in period associated with phase advance was greater, by a factor of 3 or more, than the change associated with phase delay. In computer simulations of the kind undertaken here, it would be exceedingly difficult to detect differences in period of such small magnitudes, without greatly increasing the number of pacers in the ensemble. It is, however, of some interest to determine whether effects of this sort are in principle to be expected of Coupled Stochastic Systems, granted a pacemaker with parameter values which give rise to after-effects of other sorts.

A first possible explanation for these after-effects is implicit in Figures 12.2 and 12.3. Careful consideration of these illustrations indicates that a single large, pulse-induced phase shift of discriminator activity might well induce appreciable alterations in subsequent period of the system — *provided* that the phase shift occurs at the appropriate stage during the development of the illustrated long-term trend in pacemaker period. For example, phase delay of the pacemaker rhythm early in this process (on, say, the 4th or 5th cycle shown in Fig. 12.2) ought to produce an immediate loss of some of the unentrained pacers into the inactive portion of the pacemaker cycle, and thereby cause period to lengthen during the immediately ensuing interval. That is, a single phase delay of the system ought to accelerate the "decay" process, by producing the equivalent, say, of a jump from day 5 to day 20 of the spontaneous period-lengthening process illustrated in Figure 12.2. Similarly, phase advance of the system, at a somewhat later stage − say day 10 − ought to lead to the temporary "recapture" of some of the unentrained pacers, and thereby shorten pacemaker period for the ensuing interval, relative to the period value expected, had the spontaneous behavior remained undisturbed: a jump "backward" to an earlier stage in the spontaneous process of decay in period.

Such changes in average period, which might be expected as after-effects of a phase shift, would conform with the trends observed in the experimental data, in the sense that a phase advance would produce subsequent shortening of free-running period and phase delay would produce lengthening of period. After-effects due to this process, however, would be critically dependent upon the time of phase shifting, during the spontaneous decay of the pacemaker from shorter to longer values of period, from an all-in-phase situation to a less coherent state. Of more general interest is the question of whether comparable after-effects might arise, due to single phase shift of the rhythm, when the pacers have attained an equilibrium distribution of phases.

The idealized diagrams of Figure 12.9 illustrate a process by which after-effects of phase shift should be expected even under equilibrium conditions. In the initial portions of these diagrams, discriminator activity is shown by open blocks, and the onsets of discharge by each of many unentrained pacers, all with the same basic period, are indicated by dots. The illustrated phasing of these pacers, including the regular spacing of dots during most of the inactive portion of the pacemaker cycle, is, of course, unrealistic; it does, however, indicate the relative probability that any given phasing would be realized in steady state, and therefore reflects an idealized long-term equilibrium expectation. Granted that the block of feedback is subject to a once-only advance in its timing (Fig. 12.9A), the return to an approximately equivalent steady-state distribution of phases by the pacemaker can potentially require a great many cycles. (In the absence of stochastic variation, it would, of course, never occur.) During that transition, the accumulation of pacer discharge in the hours before onset of feedback ought to lead to shortening of the ensemble period. For pacers with basic periods closer to that of the ensemble than the case illustrated, these effects would last even longer than shown. The magnitude of the effect will also, of course, depend upon the magnitude of the phase shift. Following once-only phase delay of discriminator activity (Fig. 12.9B), there would be a brief interval (two cycles in the case illustrated) during which a smaller number of discharge phases by unentrained pacers would be available to contribute to

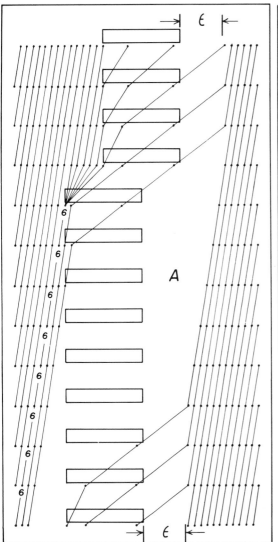

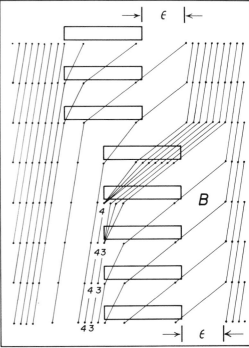

Fig. 12.9 A and B. Schematic diagrams showing the idealized expected effect of a once-only phase advance (part A) or delay (part B) of the pacemaker upon the phase distribution of unentrained pacers with basic period near to that of the ensemble. Times of discriminator feedback are indicated by elongate *open rectangles;* successive onsets of discharge by a given pacer are interconnected by *diagonal lines.* For the first several cycles, a perfect equilibrium distribution of onsets of discharge is assumed. Following phase shift of the pacemaker, there is an accumulation of pacer onsets with identical phasing; the *series of numbers* inserted into diagonal-line sequences (e.g., 6-6-6. . .) indicate the number of pacers which would begin discharge at the indicated phase. Stochastic variability would, of course, lead to a dispersal of these accumulations

onset of feedback. The expected result, then, would be a slight lengthening of the ensemble period during this interval. The magnitude of the effect in the first couple of post-stimulus cycles would depend on the magnitude of the phase shift, but the persistence of the effect would not be appreciably greater even if the basic period of the affected pacers were much closer to that of the ensemble.

Since the induced alteration in period, in the phase-delay case, would persist for appreciably fewer cycles than in the phase-advance case (compare Figs. 12.9A and 12.9B), the estimates of *average* change in period, calculated over, say, ten cycles, might well be expected to be smaller for phase delay than for phase advance — as seen in the experimental data. These considerations indicate, therefore, that Coupled Stochastic Systems ought, in principle, to conform with experimental data of Pittendrigh and Daan (1976a) not only in terms of the resultant *direction* in which period is altered by single-pulse phase shifting, but also in terms of the relative magnitudes of phase-advance and phase-delay results. Whether the models *in fact* have this capacity has not been determined by simulation. Striking results which agree with these expectations have been obtained from individual simulations, but not consistently so. The effects observed in animal experiments are extremely small on the average, and comparably small effects would not be likely to be detected in simulations of the magnitude which I have undertaken.

12.A.2 After-effects of Photoperiod

In addition to other after-effects, Pittendrigh and Daan (1976) have drawn attention to changes in period of circadian rhythms associated with the photoperiod, i.e., the number of hours of light in a 24-h cycle. Interpretation of these results is more complicated than with other sorts of after-effects, both for experimental data themselves (as Pittendrigh and Daan 1976a have emphasized) and for the pacemaker models. The central problem is the following: if an animal has a free-running period which is, say, shorter than 24 h, the process of entrainment to *any* 24-h light cycle, regardless of photoperiod, would be expected to lengthen the animal's free-running period, as an after-effect of an entrainment regime, of the sort observed both in the experimental data and in performance of the pacemaker model. (For after-effects of entrainment, see above.) An imposed 24-h light cycle can, however, exert an additional influence on the animal's behavior, which is relevant to Coupled Stochastic Systems, whenever the photoperiod is such as to alter appreciably the duration of discriminator activity, relative to its free-running value. When, for example, a 24-h light cycle includes only a few hours of darkness, a nocturnal rodent will have its daily bouts of activity "squeezed at both ends" by the light, and this compression of the daily activity may lead to opposite kinds of after-effects from those due to period of the entraining light cycle alone.

On the basis of material presented in preceding chapters, it is evident that period of the pacemaker, duration of discriminator activity, and phasing of the unentrained pacers are intimately interrelated. If the pacemaker, with nocturnal formulation of threshold, were to be entrained by a threshold cycle which is the equivalent of 6 h of light in 24 h, the resultant duration of discriminator activity would be altered to some small extent, relative to its free-running value. That entrained situation would involve attainment of an equilibrium distribution of phases for unentrained pacers, which is determined in part by the prevailing duration of discriminator activity. If, then, the period of the light cycle were kept constant, but the daily duration of "light" were increased from 6 to 18 h, this would, of course, compress the duration of discriminator activity. That would, in turn, lead to changes in the expected equilibrium phasing of the unentrained pacers.

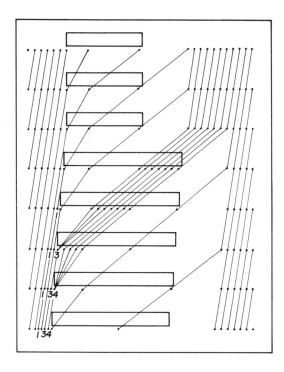

Fig. 12.10. Schematic diagram like those of Figure 12.9, to illustrate expected phasing of unentrained pacers during and following entrainment to a 24-h light regime which "compresses" duration of discriminator feedback. When the system is returned to free-running conditions (cycle 4), feedback expands to later times, resulting in a redistribution of phasing of the unentrained pacers, such that they accumulate shortly before onset of feedback, and would therefore accelerate the pacemaker. See legend to Figure 12.9 for explanation of symbols

Granted the appropriate parameters of the model, so that after-effects of other sorts are realized (such as those due to period of an entraining regime), there are intuitive grounds for expecting that photoperiod itself might alter the subsequent free-running period of the pacemaker. A schematic illustration of how this might come about is presented in Figure 12.10. The expansion of the duration of discriminator activity, following the entrainment regime, would result in an accumulation of unentrained pacers which begin to discharge just before onset of feedback. This should, then, lead to a shortening of the pacemaker period. Whether that effect would dominate over the after-effect of entrainment (which should lead to slight lengthening of pacemaker period in this case) would depend critically upon the values of several parameters in the model, which determine not only the free-running period of the pacemaker prior to entrainment, but also the duration of discriminator activity during free-running conditions. For example, if the duration of feedback under free-running conditions is only about 6 h, then an entrainment regime with more than about 18 h of light would be necessary to induce an after-effect of photoperiod, by the process envisioned here. The kind of hypothetical result shown in Figure 12.10 is in conformity with what has often been observed in experimental data (Pittendrigh and Daan 1976a), but it appears that these sorts of after-effects would be a less general property of Coupled Stochastic Systems than are other sorts of after-effects. After-effects of photoperiod *may* arise under some circumstances, but they are not necessarily to be expected, even if after-effects due to period of the entraining regime are evident.

Appendix 12.B. Alternative Steady States in a Coupled Stochastic System

The objective of the analysis here is to demonstrate the mechanism by which a Coupled Stochastic System, with entrained pacers in phase synchrony (i.e., excluding the case illustrated in Fig. 12.8), has the potential of achieving alternative steady states, with different characteristic values of period; and to examine the constraints on that possibility. Consider a pacemaker which has achieved idealized steady state with period T, in which all unentrained pacers have attained their equilibrium distribution of phases relative to discriminator feedback. Let us assume that it is a very large ensemble, so that many unentrained pacers are present with nearly identical period, pacers which might then have a phase distribution such as that shown in the upper portions of Figures 12.9A and 12.9B. Let the period of such a subset of unentrained pacers be designated as τ, and let τ be y minutes shorter than the average period of the ensemble, i.e., $T - \tau = y$.

Suppose now that due to a perturbation, some number, n, of these pacers were to be forced into an identical phasing in cycle zero, with their discharges all beginning at the start of discriminator activity, the pacers having been displaced from the time *following* discriminator activity, as in Figure 12.9A, cycle 5. If feedback were thereafter to be supposed for the moment to have no effect on these pacers in subsequent cycles, then they would free-run with their basic period, subject only to stochastic variation in basic period, to which we assign a standard deviation, s. [s is related to α, the Stochastic Coefficient of the models: $s = \alpha (1 + \delta)$.] The *average* time at which the subset of n pacers would begin to discharge would correspond to recurrences of τ; the *actual* phase of any single pacer in the jth cycle, relative to recurrences of τ, would be determined by the sum of j Gaussian variates with standard deviation s. This would lead to a distribution of onsets of pacer discharge with standard deviation which is proportional to $\sqrt{j}$, i.e., s in cycle 1, $s\sqrt{2}$ in cycle 2, $s\sqrt{3}$ in cycle 3, etc. When feedback is allowed for, the situation is of course more complex; the right-hand tail of the Gaussian distribution of phases envisioned here would be compressed, because feedback would accelerate any unentrained pacer which, by chance, did not begin to discharge until after onset of feedback by the ensemble.

Because the rise in summed pacer discharge just prior to achievement of threshold is approximately linear (cf. Fig. 5.3C), let us adopt the convention that each extra pacer, which begins to discharge shortly before onset of feedback, shortens pacemaker period by some small but predictable average amount, z minutes. At first glance, it would appear that the period of the pacemaker could be permanently shortened from T to τ, simply by fulfilling the requirement that $knz = y$, where k is the fraction of the unentrained pacer subset which begins to discharge before recurrences of τ; but an additional requirement must be met. The pacers of interest constitute that fraction of the rephased unentrained subset which begin to discharge before recurrences of τ, *after* excluding from consideration any pacer which begins its discharge so early that its discharge has ended before threshold is attained. Let us designate that fraction as K.

Now if pacemaker period T is indeed to be shortened to τ, as a result of the imposed perturbation, we require that $Knz = y$, the difference between T and τ; if this is to be a steady-state change in pacemaker period, K must converge on some nonzero value in successive cycles. It is self-evident that K in cycle 1 will be one-half; half the unentrained subset can be expected to begin discharge prior to the time specified by τ, at which time onset of feedback would occur and bring the other pacers of the subset almost immediately to discharge as well. Elementary probability calculations demonstrate that K will be about five-eights in cycle 2 (assuming that any pacer of this subset, which did not begin discharge before τ, began immediately thereafter). The expected behavior in subsequent cycles requires more elaborate calculations; among other things, we must eventually begin to exclude from consideration those pacers which begin to discharge so early that they are lost to system performance. The pacers of interest are those which begin discharging before the time specified by $j\tau$, and yet continue to discharge through that time, so as to contribute to onset of feedback. To determine those limits, one must specify the duration of pacer discharge; in approximate conformity with the parameters used in other calculations in this chapter, I assign a value of $4s$ to the duration of pacer discharge.

Simulation has demonstrated that K initially increases to a value greater than five-eights, but thereafter, in successive cycles, begins a monotonic and continual decline. The period-shortening effect of the group of unentrained pacers will eventually decay, because their discharges will eventually tend to begin so early that they can no longer contribute to onset of feedback. The spread in times of onset of pacer discharge continues to increase; the left tail of the distribution of phases moves ever farther away from onset of discriminator feedback, to the point where eventually the entire subset of n pacers would be "lost" to onset of feedback. Since K declines continually, Knz will approach zero regardless of n and z; the change in pacemaker period induced here would gradually and continually decay, and no new steady state can result.

The situation becomes a different one, however, if we assume that some fraction of the subset of rephased, unentrained pacers, $k'n$, is sufficient to accelerate the pacemaker ensemble by somewhat *more* than y, say by $y + s/5$; that is, $k'nz = y + s/5$. If steady state is to result, k' must become the fraction of n which begin to discharge before the time specified by $j(\tau - s/5)$, and yet continue to discharge through that time; let us designate that fraction as K'. We are again interested in whether K' converges in successive cycles upon a steady-state value. In cycle 1, only 42% of the relevant subset would contribute to the next onset of feedback (the integral of the Gaussian distribution from $-\infty$ to $-0.2\,\sigma$). In order to exclude from consideration any pacer which begins to discharge so early that it would be lost to the system, duration of pacer discharge must be specified. Here, as in the preceding example, it was set at $4s$. Simulation has demonstrated that K' is then about 0.52 in cycle 2, approximately 0.62 in cycle 3; and that thereafter, on up through at least cycle 100, K' remains between 0.6 and 0.7, with no clear evidence for trend. The empirical answer, then, is that K' rapidly converges, on a value of about two-thirds. Since K' converges, the phase-shifted unentrained pacers would apparently be able to induce a new steady state period in the pacemaker. To achieve this convergence, it is sufficient that $2nz/3 \geqslant y + s/5$; in this situation, the period of the pacemaker might be expected to be permanently shortened, from T to $\tau - s/5$.

There is a relatively straightforward explanation for the finding that K' converges on a steady-state value in this example, and does not in the preceding case. Without feedback, the standard deviation of the distribution of onsets of pacer discharge would increase with the square root of the number of cycles; that is, $\sigma_j = s\sqrt{j}$. If T is converted only to τ, then even with feedback, the left tail of the distribution of discharge onsets continues to expand indefinitely with increase in j. When T is converted to a value shorter than τ, however, onset of feedback advances through the distribution of pacer onsets as a linear function of j. In an approximate sense, the steady state in K' results because with increase in j, the quantity $j\,s/5$ will eventually exceed $s\sqrt{j}$. Another way of phrasing the conclusion from these calculations is that only pacers with basic period somewhat *longer* than that of the ensemble can remain fully entrained on a long-term basis; those with period identical to that of the ensemble will eventually drift away.

We should now consider whether the demands of the preceding calculation $(2nz/3 \geqslant y + s/5)$ might be achieved in a realistic situation. Consider an ensemble of 200 pacers and threshold set at 50 pacers, in which z near onset of feedback is on the order of 3 min per pacer. If nine unentrained pacers were to be rephased, so that their onset of discharge coincided with onset of feedback (i.e., $n = 9$), then shortening of pacemaker period would be about 18 min (i.e., $2nz/3$). Suppose now that $s = 60$ min; then $s/5 = 12$ min, and one might expect the new steady state to arise if the unentrained phase-shifted pacers had a period 6 min shorter than T (i.e., $2nz/3 - s/5$).

It should be noted that the process described here is to some extent self-catalyzing; while we have assumed initially that only a subset of the pacers with period τ had been rephased appropriately, the acceleration of the pacemaker would mean that eventually all other pacers with period τ would be overtaken and engulfed by the ensemble. Now it is not, of course, to be expected that nine or more pacers would exist in an ensemble of 200, all with an identical period 6 min shorter than that of the ensemble. Effects of the sort envisioned here, however, can presumably be averaged over some range of values of basic period. Bearing in mind that the population of unentrained pacers can be 50% or more of the total ensemble, the requirements for the process envisioned here do not seem to be entirely unrealistic.

Nevertheless, it is clear that the calculations described above do not constitute a rigorous demonstration that a Coupled Stochastic System can achieve alternative steady-state values of period due to phase-shifting of unentrained pacers. In demonstrating the nonconvergence of K and the convergence of K', we began with the unrealistic assumption that the subset of unentrained pacers of interest all had identical average period, and presumed further that discharge of those pacers would be immediately triggered by the onset of feedback. Cycle-to-cycle variation in period of the *pacemaker* was neglected; and no further consideration was given to a pacer which had ceased to affect feedback because its discharge occurred too early. In addition, the curve describing increase in pacer discharge before onset of feedback was assumed to have a single value of slope, unaffected by period of the pacemaker. My own intuition is that those simplifying assumptions are unimportant to the interpretation, and these calculations therefore proved enlightening to *me,* as an indication of a process by which equilibrium period value of the model could be altered by its history. It is my opinion that comparable phenomena underlie the differences in free-running period found in simulations, as illustrated in Figure 12.7. For those readers who do not share my intuition, this appendix can at best be considered suggestive about what *might* happen in the models. I regret that a more rigorous mathematical treatment of the important issue involved here requires skills beyond my own.

Chapter 13

Predictions from Coupled Stochastic Systems

13.1 Introduction

The preceding several chapters contain detailed evidence in support of
the idea that Coupled Stochastic Systems can provide a unified conceptual
framework for the interpretation of a large body of diverse experimental
data. This chapter as well as the next two are designed, in some sense, to
undo all that; a variety of predictions will be described which are derived
from such models, and which amount to ways of testing the proposed interpretations so as to identify specific difficulties, that is, to find out in what
ways the models are wrong.

The ideal method of testing models through prediction is based upon con-
trasting alternative hypotheses. One begins with a small number of conceivable explanations, each of which can account for all existing data, and
then designs critical experiments in which, if the outcome goes one way, at
least one of the extant hypotheses is excluded and can be discarded as an
adequate model; and if the outcome goes the other way, another hypothesis
or group of hypotheses is excluded. Any prediction which is fulfilled, in
terms of one hypothesis, will then also represent a mistaken prediction in
terms of an alternative interpretation. Each properly chosen experiment
thereby serves to disprove an hypothesis and thus to reduce uncertainty
(Platt 1964).

Unfortunately, the study of circadian pacemakers has not yet progressed
to the stage where even two clearcut alternative hypotheses are available.
In this sort of situation, the role of prediction is somewhat less decisive.
Prediction remains an interesting and a useful thing to attempt, but one
must avoid the inviting trap of thinking that "successful" prediction is a
desirable outcome, a demonstration that one does indeed "understand" the
system. In fact, predictions which are fulfilled are apt to be far less informative than those which prove to be mistaken. Fulfillment of any of the predictions in this chapter, for example, would contribute only a single bit of
additional evidence to the sum of that in the preceding chapters, one more
kind of experimental observation that can be accommodated naturally by
Coupled Stochastic Systems.

The set of other, fundamentally different models which might account
for that same single predicted result covers a very broad and largely un-

defined spectrum, so the new observation would do little to reduce uncertainty. On the other hand, any prediction that is shown by experiment to be wrong provides a definitive conclusion; at the least, it demonstrates that some critical feature of the animal's pacemaker has been omitted from the model. When confronted with evidence that contradicts a prediction, one can, of course, attempt to rescue the proposed model by some sort of elaboration on the initial proposal. Failure of certain key predictions offered here, however, would expose fundamental weaknesses of the entire conceptual scheme, and should stimulate the search for an entirely new starting point for the unified interpretation of *all* the available data, including the new, discordant observation. It is in that spirit that the predictions of this chapter are offered: not in the fatuous hope that correct predictions could demonstrate the validity of the class of models considered here, but in the confident expectation that at least some of the predictions are wrong. Those wrong predictions will best serve as a guide for the generation of alternative hypotheses.

It deserves emphasis that, as far as I am aware, the predictions of this chapter do not — except where indicated — involve experiments which have already been done. Instead, these are suggestions for future research. In my opinion, all the experiments proposed here are eminently feasible. No new techniques, unusual equipment or special skills are required. The experiments are in most cases so elementary in design that they could have easily been undertaken 15 years ago in any of several well equipped laboratories. Far more complex experiments of similar sorts have already been undertaken (e.g., Daan and Pittendrigh 1976a, 1976b). Most of the chapter is devoted to qualitative predictions involving aspects of the models' behavior which do not depend critically upon particular value of the parameters. In such cases, the interpretation of the experimental outcome would involve evaluating a simple dichotomy: either the qualitatively predicted phenomenon is observed, or it is not. Granted judicious choice of experimental organisms, i.e., those with relatively precise rhythms, an ambiguous result appears very unlikely to me. The final portion of the chapter is devoted to quantitative predictions of a sort which would require a fair amount of preliminary experimentation, so as to select parameter values for that model appropriate for detailed prediction. The justification for such effort lies in the diversity of predictions which can be made, once an adequate model has been selected. The intent there is to predict the consequences of exposing an animal to any conceivable light regime.

13.2 Predictions Related to Aschoff's Rule

Nearly all combinations of parameter values chosen for a Coupled Stochastic System lead to the expectation that increases in discriminator threshold should lengthen the period of the pacemaker. Coupled with the assumption that light raises threshold for nocturnal animals and lowers threshold for diurnal animals, this interpretation leads to Aschoff's rule, which was a central topic of Chapter 7. The few cases of experimental data which demonstrate violations of that rule therefore demand special attention. Three such exceptions were described in Appendix 7.A: data from an owl, a monkey, and a squirrel. The observations from the barn owl (Erkert 1969) and the rhesus monkey (Martinez 1972) seem most easily interpreted by the supplementary hypothesis that discriminator threshold of these species has an unusual, nonmonotonic relationship with light intensity: that minimum values of discriminator threshold are produced by an intermediate light intensity (see App. 7.A). The prediction that follows directly and unequivocally from that interpretation is that the normal phase relationship of the animals to a light-dark cycle (nocturnal for the owl, diurnal for the monkey) should be completely reversible by the appropriate choice of light intensities for the entrainment regime. The barn owl should become "diurnal" in a circadian light regime in which darkness is paired with a daytime light intensity of about 0.5 lux; the rhesus monkey should become "nocturnal" in a lighting regime in which light of about 5 lux serves as nighttime intensity, and much brighter light, say 1000 lux, serves as daytime intensity. Should future experimentation contradict this expectation, the proposed explanation for the existing data on free-running rhythms of these species would be mistaken. No simple remedy for such a deficiency of the models occurs to me. A failure of this prediction — granted the use of individual animals of the species involved, which show the sorts of behavior documented by Erkert (1969) and Martinez (1972) — would therefore be a very serious blow for the models, which might lead to the development of fundamentally different alternative models.

The other clearcut violation of Aschoff's rule is contained in data from the palm squirrel (Pohl 1972), and the nature of those data suggests that within the framework of Coupled Stochastic Systems, only a pacemaker in which $\bar{\delta}$ is greater than 0.5, and in which ϵ is some 10 h or more, can adequately account for this case. (See App. 7.A.) Such parameter values lead to the prediction that a several-hour pulse of bright light in the middle of the squirrel's normal sleep-time should produce a phase shift of many hours in the animal's rhythm. That experiment has not yet been undertaken, but further consideration indicates that the prediction involved here is not a particularly critical one. A failure of the prediction might conceivably only reflect severe restrictions on the extent to which light intensity can lower discriminator threshold. A more critical prediction from these values for the parameters $\bar{\delta}$ and ϵ, a prediction which is independent of such possible limitations on threshold, is that entrainment of the palm squirrel by a non-24 h light-dark cycle should proceed much more readily with entraining regimes which have periods shorter than free run, than with those which have longer. For example, if the free-running period of a given animal in very dim light is, say, 24.5 h, and entrainment by light cycles is possible only up to period values of 26 h ($\Delta\tau = 1.5$ h), then entrainment should be readily achieved by light cycles with periods appreciably less than 23 h ($\Delta\tau > 1.5$ h).

Should future experiments demonstrate that this prediction is wrong, the models would be left without adequate explanation for a small but significant body of experimental data. This would be a critical failure of the models, one which would force the conclusion that circadian pacemakers − or at least those of the palm squirrel − must incorporate some important component which is neglected in Coupled Stochastic Systems.

13.3 Predictions Relating to Entrainment by Sinusoidal Light Cycles

A series of simulations was undertaken in which 24-h sinusoidal orscillations of threshold were imposed upon a Coupled Stochastic System, so as to mimic the effects of sinusoidally varying light cycles. The parameters assigned to the model were those of the type model, except that free-running period of the pacemaker was altered by varying $\bar{X}_b$ between 16 and 18 h; and amplitude of the imposed threshold oscillation was varied from 0.05N to 0.2N. Figure 13.1 presents a graph of the observed phase relationships between onset of discriminator activity and threshold minimum (minimum light intensity for a nocturnal animal, maximum intensity for a diurnal animal). Accessory simulations indicate that qualitatively similar results are obtained even using models with parameter values quite different from those of the type model. The qualitative trends shown in Figure 13.1 therefore represent very general responses for a Coupled Stochastic System.

The following predictions follow directly from Figure 13.1: with a sinusoidal circadian light cycle of a given period and amplitude, individual animals with short free-running period are predicted to awaken earlier relative to the light cycle than those with long period; the relationship between times of awakening and free-running period should be approximately linear; and the smaller the amplitude of the entraining cycle, the greater the change in predicted phasing for unit difference in free-running period.

When one carefully considers these simulations, it is evident that there is an intrinsic interchangeability between free-running period of the animal's

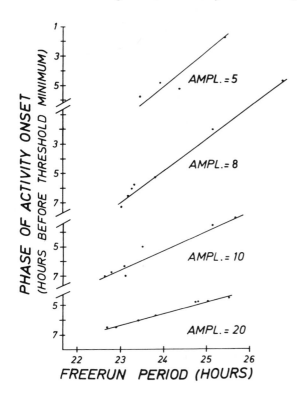

Fig. 13.1. Phase relationship observed at steady state from a series of simulations with sinusoidal oscillations in threshold, the amplitude of which (AMPL., expressed in number of pacers) was varied. Free-running period of the pacemaker ensembles was systematically varied by assigning to $\overline{X}_b$ values from 16 to 18 h, while all other parameters had values as in the type model; 750-h simulations, with N = 100 pacers. See text for predictions implicit in these results

rhythm and period of the entraining light cycle. Entrainment of a pacemaker which has a 25-h free-running period by a threshold cycle with 24-h period would be essentially equivalent to entrainment of a pacemaker with 24-h period by a threshold cycle with 23-h period; the difference is only one of scaling. Hence, three additional predictions can be indirectly derived from the simulations results of Figure 13.1: for an animal with a given free-running period, and a sinusoidal light cycle with given amplitude, times of awakening should be earlier in a light cycle with long period than in cycles with short period; the relationship between time of awakening and period of the light cycle should be approximately linear; and the smaller the amplitude of the entraining light cycle, the greater the expected change in phasing predicted for unit difference in entraining period.

The phenomena involved in these predictions are by no means peculiar properties of Couples Stochastic Systems; roughly similar predictions follow from a wide variety of basically unrelated models, including even many models for elementary oscillatory systems which are obviously irrelevant to circadian rhythms. These predictions are not, however, a completely empty exercise, because of the predicted *linearity* between period of the rhythm and phase relationship, which applies specifically to sinusoidal threshold regimes. In simulations in which the entraining regime is formulated as a square-wave cycle in threshold, Coupled Stochastic Systems predict striking departures from linearity, in the relationship between period of a circadian rhythm and phase relationship during entrainment (Chap. 9; see also below). As noted earlier, the available experimental data from both nocturnal and diurnal animals conform with those predicted sorts of nonlinearity. The central interest of the predictions in Figure 13.1, therefore, is in the implication that the documented nonlinearities in phase relationship with square-wave light regimes are a product of the kind of light cycle administered, and are not fundamental peculiarities of the circadian wake-sleep cycle.

13.4 Predictions About "Phasic" Effects of Light

For both nocturnal rodents and diurnal birds, certain aspects of the available experimental data required the assumption that a step-up in discriminator threshold, produced by a sudden change in light intensity (increase for nocturnal rodents, decrease for diurnal birds) be associated with a transitory overshoot effect (cf. Figs. 10.7 and 11.2). None of the available experimental data, however, demanded the reciprocal assumption that a step-down in threshold produce a transient supplementary excitatory effect; and no such effect was incorporated into the models.

In order to test this interpretation, consider the following experiment: a bird is placed under continuous light sufficiently bright to cause arrhythmia, and is then subjected to a supplementary lighting regime, in which intensity is very gradually increased over several hours, and then suddenly switched off. The overshoot effect expected, due to switching off the lights, is predicted, according to the assumption illustrated in Figure 11.2, to increase threshold to the extent that feedback would be temporarily interrupted, and it should therefore be possible to entrain the rhythm of the bird by regular circadian repetitions of this kind of sawtooth lighting regime. It should not, however, be possible to entrain the bird's rhythm with a converse light regime, in which supplementary light is suddenly switched on, and then very gradually dimmed to the original level.

These experiments have not been done, although it should be noted that a somewhat related experiment has been undertaken with the house finch. I have found that the circadian rhythm of a bird can be entrained with a square-wave lighting cycle superimposed upon a background light intensity, which is sufficiently bright that it leads to arrhythmia, when provided continuously (Enright unpublished). That experiment does not distinguish, however, as the above prediction does, between possible tonic action of the supplementary light itself, and the action of either or both of the sudden transitions in intensity.

As a simple extension of the "sawtooth lighting" experiments proposed here, the prediction also follows naturally from the assumptions of the models that when a bird becomes arrhythmic due to sufficiently bright constant light, a supplementary lighting regime with very gradual changes in intensity (e.g., a sinusoidal light cycle) should be unable to entrain the bird's rhythm no matter how bright the supplementary light is made.

Should it prove possible to entrain the bird's rhythm in these experiments with a step-up plus gradual decrease in light intensity, further elaboration of the models would be essential, an additional complexity which is apparently required by none of the existing data. Should it prove possible to entrain the bird's rhythm in these experiments by means of sinusoidal light cycles, one would, I believe, be forced to invoke some kind of continuous relationship between summed pacer discharge and discriminator feedback, such as those illustrated in Figure 8.1B, thereby demanding the added complexity of at least two otherwise unneeded parameters to describe the response of the models.

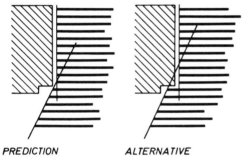

PREDICTION   ALTERNATIVE

Fig. 13.2. Qualitative prediction for outcome of experiment with nocturnal animal: following premature shutoff of the light regime (*crosshatched*) on the final day of entrainment, the rhythm is predicted to show only its free-running period (i.e., without a phase shift, as indicated in the alternative). Should this alternative result be obtained, it would imply a phasic stimulatory effect of sudden light transitions which lower threshold: an assumption not otherwise required

In order to test for the possibility that the sudden offset of light induces a mirror-image overshoot of threshold in nocturnal animals — noting, again, that the models do not incorporate such effects — consider the experiment illustrated in Figure 13.2. Should sudden decreases in light intensity produce a supplementary, transient effect on discriminator threshold, a phase shift of the nocturnal animal's rhythm, toward an earlier activity onset, would be predicted due to the final light-out stimulus. Coupled Stochastic Systems do not make provision for such a phase shift. Since the difference in timing between the two alternatives would be a relatively small one in this case, extensive replication should be incorporated into the experimental design, so as to avoid ambiguity in the interpretation.

13.5 Predictions About Transients

As described in detail in Chap. 10, experiments with nocturnal rodents generally have shown that phase advance of the animal's wheel-running rhythm due to a single light pulse is accompanied by several transient cycles before the steady-state period is attained (cf. Figs. 10.1 and 10.2). Coupled Stochastic Systems can show comparable behavior (Fig. 10.9, parts C and D). In an insect, qualitatively similar transient cycles have been interpreted (Pittendrigh 1965) to mean that a light-sensitive pacemaker is immediately and completely shifted by the stimulus to its new steady-state phase, but that the monitored function — eclosion in the flies, perhaps locomotor activity in the present context — is a rhythmic process which is temporarily uncoupled from the light-sensitive pacemaker, and which gradually returns, during the transient cycles, to its stable phase relationship. In Coupled Stochastic Systems, however, the sensitivity of the pacemaker to light is directly linked with feedback from the discriminator, and hence, with locomotor activity. No flexibility in the phase relationship between light-sensitivity and the activity pattern is envisioned, no form of intermediate coupling as proposed for the flies. The transients of Figure 10.9 correspond directly to the phase of the light-sensitive pacemaker of the system. The models predict, therefore, that during the transient cycles fol-

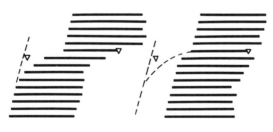

PREDICTION   ALTERNATIVE

Fig. 13.3. Qualitative prediction for the outcome of an experiment relating to the meaning of transient cycles during phase advance of the rhythm of a nocturnal rodent. A single light pulse is administered at a time which is expected to lead to several advancing transients before the new steady state, and a second light pulse is administered about an hour before expected onset of activity. The prediction is that the second pulse will be without effect, so that the full expected phase advance will be realized; the alternative (right), which is not compatible with the models, is that the light-sensitivity of the pacemaker system is fully phase-shifted on the first day, so that the second light stimulus would induce a delay relative to expectations without that second stimulus. See text

lowing an advancing phase shift of a nocturnal rodent, the normal insensi-
tivity to light should prevail during the entire inactive time. Figure 13.3
presents in schematic form an experiment by which these alternatives can
be distinguished.

A failure of this prediction — and the experiment appears to me to be a
relatively simply one — would represent a crippling blow for the models.
It would indicate, at the very least, that one of the more significant accomplishments of Coupled Stochastic Systems, the interpretation provided for
the occurrence of transients with phase advance, and none with phase delay,
is seriously in error. It would furthermore raise doubts about the mechanisms
proposed here to account for phase advance of the circadian rhythms of
nocturnal rodents, and would invalidate the general interpretation that
locomotor activity is ordinarily an adequate measure of phase for the light-
sensitivity of the circadian pacemaker.

13.6 Prediction About After-effects

Among the after-effects documented in the circadian rhythms of nocturnal rodents, the least conspicuous is the slight tendency for an animal's
rhythm to have a longer period following single light pulses which delay the
rhythm. While there is indeed a trend in several species for after-effects of
this sort, the magnitude of the reported difference in period is quite small,
with average values of less than 3 min reported for three species of mice
(Pittendrigh and Daan 1976a). In the evaluation in Chap. 12 of how after-
effects due to phase shifting might arise in Coupled Stochastic Systems,
the first possibility suggested was that the phase shift might either accelerate
or retard the spontaneous decay of after-effects which have some other
origin. This interpretation leads to the following prediction: the after-effects
(increase in period) due to single-pulse delay of a nocturnal animal's rhythm
should be enhanced, if the delaying stimulus is given at a time at which the
rhythm has an unusually short period induced by prior treatment. For example, when a mouse is entrained to, say, a 21-h light cycle, and then returned to constant darkness, an unusually short free-running period is to be expected, which will gradually increase to a more normal value. If a single
light stimulus is then administered within a few days after such an animal
has been returned to darkness, at a phase which produces large phase delay
of the animal's rhythm, one should expect both a single-step increase in
period of the rhythm, and a decrease in the rate at which period changes
thereafter. This prediction is illustrated in schematic form in Figure 13.4.
The critical issue is not simply that a change in period should be inducible
by phase delay, since very small trends in this direction have been previous-

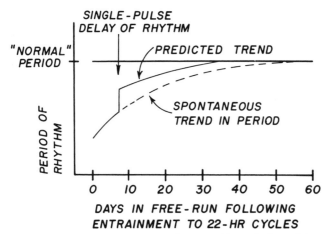

Fig. 13.4. Qualitative prediction for the effect of phase delay on the period of a nocturnal rodent's rhythm, during the first few days after release into constant conditions, following prior entrainment to a short-period lighting regime. See text for rationale underlying this prediction

ly documented. The prediction here is that the pretreatment should increase the magnitude of the change, so that the increase in period becomes a clear-cut, reliable phenomenon. Furthermore, the resulting change in period should be roughly proportional to the magnitude of the induced phase delay of the rhythm.

A failure of this prediction would raise serious doubts about the validity of the interpretation offered by Coupled Stochastic Systems for after-effects. In my opinion, that interpretation is one of their most interesting and important merits. After-effects are by no means the most conspicuous characteristic of the circadian wake-sleep rhythm, but subtle as they are, I regard them as an important feature of experimental data, and one for which a satisfactory mechanistic explanation has not previously been available. If simulations based on Coupled Stochastic Systems have offered the wrong explanation for after-effects, the entire substance of Chap. 12 has been an empty exercise, describing how the pacemakers *might* function, but in fact do not.

13.7 Predictions About Phase Shifting by "Dark Pulses"

According to the assumptions made in Chap. 7, the primary qualitative difference between diurnal and nocturnal animals lies in the opposite effects of light upon threshold of the discriminator. In principle, then, a dark pulse administered to a diurnal animal should be the equivalent of a light pulse administered to a nocturnal animal. If, therefore, a diurnal bird were to be placed under a constant level of moderate light intensity, leading to a free-running rhythm, and the bird were then subjected to brief intervals of darkness, the resulting phase-response curve can be qualitatively predicted. In two features, it should resemble that of nocturnal rodents (Fig. 10.3): dark treatment at or shortly after activity onset should lead to a modest phase delay of the rhythm; and dark treatment in the middle of the bird's sleep should have no effect on phase of the rhythm. There, however, the predicted resemblance ends. As was demonstrated in Chap. 11, the values for the parameters $\bar{\delta}$ and ϵ required to reproduce the phase-response curve of the sparrow are somewhat smaller than those appropriate for the pacemakers of nocturnal rodents. With such values for these parameters, it should not be possible to produce phase advance of the rhythm of a nocturnal rodent by light pulses; nor should phase advance be possible when single dark pulses are administered to a bird. The entire predicted phase response curve, therefore, should resemble the delay portion of that of a nocturnal rodent, with complete unresponsiveness, not only during the inactivity time, but also at the time at which phase advance is obtained with the rodent.

The comparable prediction, for dark treatment of a nocturnal animal, would require maintaining the animal at some relatively weak level of light intensity, such that wheel-running rhythmicity continues. Since darkness is here regarded as being less stimulatory for a nocturnal animal than light is for a diurnal animal, a dark pulse cannot be expected to lower threshold to as great an extent as a light pulse would for a diurnal bird. That restriction has a number of qualitative consequences, leading to the prediction of very modest phase advance when a dark pulse is given just before activity onset; slight phase delay when a dark pulse is given near the expected end of locomotor activity; and no effect either during wakefulness or during the majority of the sleeping time.

A failure of these predictions would undermine critical assumptions of Coupled Stochastic Systems which are involved in all interpretations of the influences of light on the pacemaker. In particular, I see no way in which large phase shifts, either advances or delays, might be induced by pulses of darkness without demanding radical revision of many of the conclusions and interpretations of preceding chapters.

13.8 Quantitative Predictions

In order to make detailed quantitative predictions, it is necessary to select values for a full suite of parameters for a Coupled Stochastic System, values which are appropriate to the species, or, better yet, to the individual animal to be tested. The selection of these values can probably best be guided by iterative attempts to fit the data from a carefully measured phase response curve. It was demonstrated in Chapters 10 and 11 that phase-response-curve data place severe constraints on acceptable parameter values. In addition, one should examine the dependence of free-running properties of the animal's rhythm, both period and duration of activity, upon light intensity during constant conditions. This latter step is essential for the calibration of discriminator threshold against light intensity, and it will also serve as a check on parameter values chosen for the model. It is entirely possible that a given phase response curve might be simulated by several rather different pacemakers, with alternative sets of values for the various parameters, but it seems highly unlikely to me that both phase-response-curve data and data from the free-running rhythm at several light intensities would leave appreciable ambiguity in the selection of parameter values.

Once the parameters have been evaluated by the appropriate preliminary experiments, there is no limit to the variety of quantitative predictions that can be made by the model, involving new varieties of light treatment. It is evident in Figure 11.6 that the model used to derive the phase response curve for a diurnal bird led to only a moderately good fit to the experimental data for sparrows. Further simulations could, no doubt, permit more refined choices of parameter values, so as to improve the correspondence. The following quantitative predictions have been derived from simulations which used the parameter values leading to Figure 11.6A; they should therefore be regarded as only approximate: examples of the kinds of use to which a completely evaluated model can be put, rather than definitive predictions for sparrows.

With this qualification, one could, as a starting point, ask how the entrained phase relationship of the bird's rhythm should vary as the period of a square-wave light cycle (incorporating 6-h light stimuli) is varied. Results from an array of simulations addressed to this question are presented in Figure 13.5. It turns out that this particular exercise is not prediction in the same sense otherwise used in this chapter, since some of the experiments simulated here have been undertaken with birds. The data from comparable experiments with sparrows (Eskin 1971) are also presented in Figure 13.5. The correspondence between prediction and observation is by no means perfect; note, particularly, the discrepancy of about 3 h in phase at a period value of 21.3 h. That discrepancy may, however, reflect a complication in evaluating activity onset in the bird experiments; in the simulations, time of onset was determined by examining phase at which subsequent free run began. In any case, the fit to experimental observation appears to be as good as the agreement between simulation and the original phase-shift data (Fig. 11.6B).

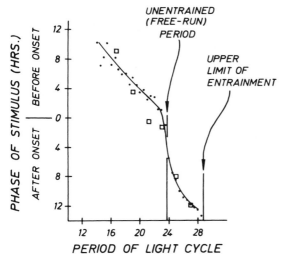

Fig. 13.5. Phase relationship between onset of activity and 6-h light stimulus, as a function of period of the cycle. *Solid points*: computer simulations based on a model with parameters like those which led to the phase-response curve of Figure 11.6A; *open squares*: experimental data for sparrows (Eskin 1971). For light regimes with period shorter than free-running, phase was determined by first activity onset after end of entrainment (except for sparrow data with 21-h cycle). The free-running period of the sparrow rhythms was about 25 h; comparisons with the simulations (free-running period of about 24 h) therefore required subtraction of an hour from the periods of the light regimes, since the relevant variable is difference between free-running period and period of the entraining cycle

The computer simulations did not, however, show any clear lower limit to the range of period values permitting entrainment, although Eskin (1971) reported (without illustrative examples) a loss of entrainment by three sparrows out of four tested in light cycles with period of 15.8 h. This may be a genuine discrepancy between behavior of the model and that of the birds; on the other hand, because both model and bird show forced activity during the light phase of the imposed regime, it would, I think, be very difficult, with bird or model, to distinguish, even in terms of subsequent free run, between true entrainment with such a lighting regime, and the exogenous effects of light upon locomotor activity.

As a further example of the questions which can be addressed by computer simulations with the model, one can determine the kind of phase-response curve expected if the duration of the light stimulus is varied. Figure 13.6 presents data from simulations in which "light" stimuli with durations of 1, 3, 9 and 12 h were administered, together with two sets of results for 6-h stimuli, one of which was illustrated previously in Figure 11.6A. One can also ask what kind of phase-response curve should be obtained for 6-h stimuli, when the background light intensity is increased (Fig. 13.7).

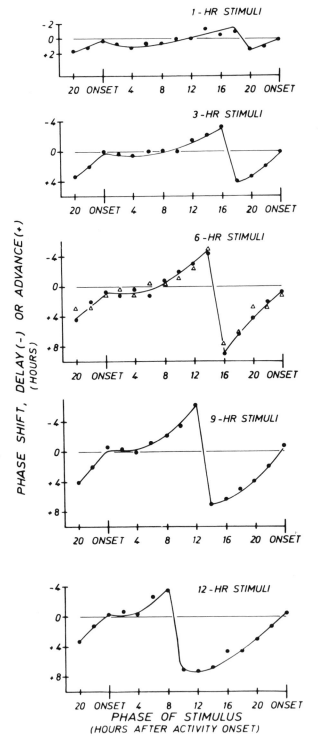

Fig. 13.6. Phase-response curves obtained in simulations with models like that which led to Figure 11.6, in which the light stimulus was of 1-h, 3-h, 6-h, 9-h and 12-h durations. The *solid points* in the curve for 6-h stimuli represent the data of Figure 11.6; the *triangles* are from another comparable series of simulations, with another ensemble of pacers (different values of $z_{i,\beta}$ and $z_{i,\gamma}$)

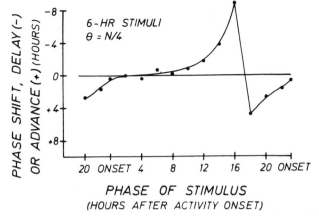

Fig. 13.7. Phase-response curve obtained in simulations with the model used for the results of Figure 11.6 and 13.6, but with threshold of 0.25N instead of 0.3N, with the result that the free-running period was somewhat shortened. A comparison of this curve with that of Figure 13.6 for 6-h stimuli indicates that there are large-magnitude differences in the predicted amount of phase shift, and even in direction of shift, due to this small difference in threshold

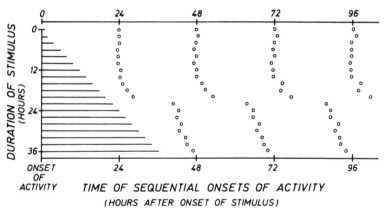

Fig. 13.8. Sequential onsets of discriminator activity from simulations with the model which led to the phase-response curve of Figure 11.6, in which single stimuli of 2, 4, 6, . . . 36 h duration were administered beginning at activity onset, in a free-running rhythm. Each simulation result constitutes a single horizontal row of data points

One can ask how the animal should respond to single-light stimuli of different durations, all beginning at a given phase of the rhythm, such as activity onset (Fig. 13.8). One can ask how the entrained phase should depend upon the background light intensity provided during "darkness" (Fig. 13.9).

Experiments with the house finch (Hamner and Enright 1967) have demonstrated that it is possible to entrain the rhythm of the bird to an aver-

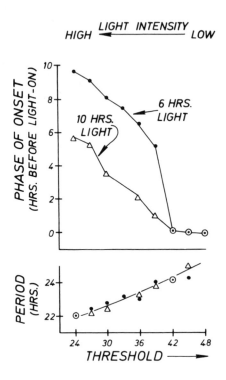

Fig. 13.9. Steady-state phase relationships predicted under various light regimes, obtained in simulations with a model like that used for Figure 11.6, except that $\overline{X}_b$ was assigned a value of 17 rather than 18 h. Differences in threshold represent differences in background light intensities, upon which a 24-h cycle of bright light (i.e., lowered threshold) was superimposed, with either 6 or 10 h of light per cycle. Threshold values represent percentages of N. The *lower portion of the figure* shows the free-running period values of the rhythms which resulted at these background light intensities. Note that a difference in light intensity which changes free-running period by 2 h (threshold increase from 0.24N to 0.42N) changes the entrained phase relationship by about 10 hours, for a 6:18 light-dark regime, and by about 6 h for a 10:14 light-dark regime

age 24-h period by means of a 36-h light cycle, with 6 h of bright light and 30 h of dim light. The light stimuli alternately produced phase delay and phase advance of the rhythm, leading to a repeating pattern every 72 h. Similar results were obtained using a 60-h light regime, with 6 h of light and 54 h of darkness. A rhythm with *average* period of 24 h resulted; the activity pattern repeated itself every 5 days. Given these results as background, a properly calibrated model could be used to predict the range of period values of the light cycle (around 36 h and around 60 h) with which this kind of entrainment should be achievable. Other examples of the kinds of questions, for which I have not performed the required simulations, but for which a calibrated model could make quantitative predictions, include the following: How does the phase response curve change as the intensity of the stimulus is decreased from the very bright level of light assumed in previous simulations? How should the animal respond to two or more light stimuli per circadian cycle? How should the animal respond to a combination of light pulses and dark pulses? The specific predictions and questions posed here are by no means an exhaustive list, but they should be sufficient to suggest the range of possibilities. Once the parameters of the model have been appropriately evaluated in preliminary experiments, quantitative predictions can be made for the expected consequences of any conceivable lighting regime.

The most productive use of this sort of prediction would be to undertake arrays of simulations from many hypothetical experiments and to sift
through the results in the search for "surprising" predictions, either those
which seem intuitively unlikely or those which clearly conflict with expectations from some other sort of model. It is the experimental test of the surprising predictions which will provide the best index of the extent to which
the models are adequate to account for the behavior of the animals.

Chapter 14

Further Predictions: a Modest Success and Two Problem Cases

14.1 "Clamped" Free-run Experiments

One of the essential features of Coupled Stochastic System is that the value of discriminator threshold prevailing at the time the animal awakens plays a critical role in pacemaker performance. Regardless of parameters of the specific model, modifying the value of light intensity prevailing at this time should appreciably alter period of the pacemaker, not only on a single-cycle basis as in phase-shifting experiments (preceding chapter) but on a long-term basis as well. In order to examine this expectation while minimizing other effects of light on the system, a novel experimental design is proposed here, in which a circadian light cycle is administered so that the timing of a light stimulus is "clamped" to some fixed phase of the pacemaker rhythm. A modified or clamped free-run will result, with a period determined by the responses of the pacemaker to the repetitive light stimuli.

Such an experiment can be undertaken as follows: the switch on the running wheel of a nocturnal rodent is attached to a counter, which is connected to a relay, so that as soon as, say, a few dozen wheel revolutions have been completed, the prevailing lights go out and stay out for, say, 18 h. At the end of that interval (and while the animal can be expected to be asleep), the lights then come on automatically and remain on until again extinguished by the animal's activity in the running wheel. Under these circumstances of clamped free run, the period of the animal's rhythm is predicted by the models to vary markedly, depending upon the light intensity provided. That light intensity would determine the level of discriminator threshold prevailing at the time of onset of feedback; for a nocturnal animal, an increase in intensity is predicted therefore to increase the period of the rhythm. Bearing in mind that a Coupled Stochastic System is completely insensitive to stimuli which raise threshold during most of the inactive time, this effect of light intensity is predicted to be completely independent of whether the light is switched on 14 or 22 h after activity onset.

Were the light intensity which is used in such an experiment to be provided on a continual basis, the light would, according to the models, be expected to terminate feedback after a shorter interval than would be expected during free run in darkness, and that ought to have a weak tendency to shorten period of the rhythm. In the proposed clamped free-run experiment, however, the animal would fall asleep in darkness; it would receive the full period-lengthening effect of light just before onset of activity, without the counteracting effect on end of feedback. Hence, the influence of light intensity upon period of a clamped free-running rhythm is predicted to be greater than the effect of the same intensity during ordinary free run.

A similar experiment can also be undertaken with a diurnal bird. The perch of the cage can be provided with a counter, connected to a relay, so that as soon as the bird has hopped onto and off the perch a few dozen times, bright lights would be switched on and left on for, say, 8 h. The duration of the bird's locomotor activity would, of course, be determined by the duration of bright lights. Coupled Stochastic Systems predict that changing the value of the continuous "background" illumination, which would prevail before activity onset, should have even stronger effects on period of the bird's rhythm than an equivalent change in intensity of continuous illumination under ordinary free run. For example, if an increase in continuous light intensity from 0.01 to 0.1 lux shortens free-running period by 30 min, then in the experiment proposed here, the difference in clamped free-running period associated with these same values of light intensity should be somewhat greater than 30 min.

Consider now an alternative in this latter experiment, in which the background light intensity at which the bird awakens is held constant, and the *duration* of the bright lights is systematically manipulated, thereby altering the duration of the bird's wakefulness. Long duration of light can, of course, increase duration of wakefulness over its free-run value; and if the lights are suddenly switched off, this can prematurely terminate locomotor activity and shorten wakefulness relative to the ordinary free-running value, in keeping with the "overshoot" assumption of Figure 11.2.

As a preliminary to this proposed experiment, the free-running period of the bird's rhythm should be measured under the light intensity chosen for background conditions, as well as the duration of locomotor activity associated with that period. The prediction which follows from a broad class of Coupled Stochastic Systems is the following: if duration of the bright lights, provided by the switching circuit, is less than the duration of wakefulness under free-running conditions, the period of the bird's clamped free-running rhythm should be shorter than that observed during ordinary free run; and with increasing duration of bright light, the period of the bird's rhythm should increase monotonically.

I have conducted one experiment of the latter sort with a single individual male house finch, and the results during some 6 months of the experiment are illustrated in Figure 14.1 and summarized in detail in Table 14.1. Minimum values of period of the rhythm were obtained with 10 h of light; each increase in duration of light led to lengthening of period, each shortening

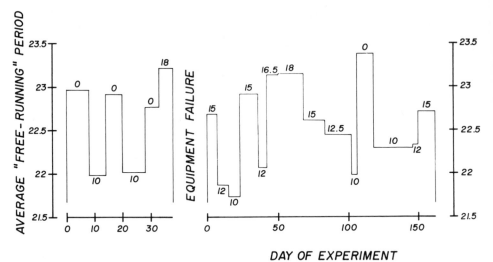

Fig. 14.1. Average period values observed of a house finch during a clamped free-running experiment, in which duration of lighting (indicated by numbers adjacent to bars in the graph) was varied. A duration of "0" represents simple free run, without supplementary lighting; other values expressed in hours. Under each nonzero condition, the bright lighting, of predetermined duration, was initiated automatically in each cycle by the bird's spontaneous activity, and began about 15 min (range: 10 to 25 min) after activity onset. Light intensities: about 0.1 lux as background, about 200 lux as imposed supplement. See text for interpretation

to shortening of period: a qualitative confirmation of prediction. During the first three free-running experiments ("0" duration of lights), the duration of the bird's locomotor activity had average values between 12 and 12.5 h (range: 10.5 to 15 h). Hence, it was qualitatively expected, as observed, that 10 h of bright light would shorten the period, relative to free run, and that 18 h would lengthen the period relative to free run. While this qualitative effect can be predicted on the basis of general considerations, a fully specified model, with all parameters determined (cf. Chap. 13) would be necessary to establish properly whether the magnitude of the observed changes in period conform with expectations for a Coupled Stochastic System.

In the only free-run test during the second portion of the experiment, several months later, the free-running period of the bird's rhythm had lengthened by about 30 min, and the duration of activity under free run had increased to an average of more than 15 h. In this case, the prediction is that light of less than 15 h duration should shorten period of the rhythm, relative to free run. In view, however, of the non-stationarity in both period of the rhythm and duration of activity during free run (i.e., the change from about 12 to about 15 h activity), as well as the long interval between tests of free-running properties, this quantitative comparison of free-running period with

Table 14.1. Results from clamped free-run experiment

Number of days of treatment	Light condition (h/cycle)	Mean Period (h:min)	Standard deviation (min)	Standard error (min)	Range of period values (h:min)	Mean duration of activity (h:min)
8	0	22:58	11.1	3.9	22:47–23:15	12:19
6	10	21:59[a]	13.1	5.4	21:40–22:10	
6	0	22:55[a]	10.2	4.2	22:43–23:13	12:01
8	10	22:01[a]	17.9	6.3	21:25–22:26	
5	0	22:46[a]	12.0	5.3	22:32–23:03	12:26
5	18	23:13[a]	7.4	3.3	23:08–23:26	
		Equipment Failure				
7	15	22:41	12.6	4.8	22:22–22:58	
8	12	21:52[a]	7.6	2.7	21:42–22:05	
8	10	21:44	17.7	6.3	21:20–22:02	
13	15	22:55[a]	26.8	7.4	22:23–24:07	
6	12	22:04[a]	7.5	3.1	21:53–22.13	
8	16.5	23:08[a]	27.8	9.8	22:36–24:09	
18	18	23:09	11.1	2.9	22:49–23:31	
15	15	22:37[a]	14.3	3.7	22:07–22:53	
19	12.5	22:27[a]	9.6	2.2	22:10–22:40	
4	10	22:04[a]	2.2	1.1	22:04–22:12	
12	0	23:23[a]	23.0	6.7	22:55–24:27	15:18
28	10	22:18[a]	12.6	2.4	21:47–22:40	
4	12	22:20	5.1	2.5	22:14–22:26	
12	15.4	22:43[a]	9.4	2.7	22:26–22:58	

a Mean period value significantly different from that under preceding condition.

that under various clamped free-run conditions involves a large measure of uncertainty. It is not clear whether the difference between the value of free-running period (23:23) and those with 15 h of light (22:41, 22:55, 22:37 and 22:43) is indeed a serious discrepancy, although it appears to be so. Clearly, in a repeat of this experiment, it would be essential to make "control" observations of unclamped free run at regular intervals, as was done in the initial part of the experiment, so as to permit more rigorous comparison with expectation. A conservative interpretation of these results, then, is that they agree qualitatively with prediction (longer light consistently led to greater period), with some uncertainty about the quantitative aspect (the relationship of duration of bright light for a given period to the duration of wakefulness under free-running conditions).

The original intent of the experiment, shown in Figure 14.1, was to contrast the predictions of a Coupled Stochastic System with those of a generalized mathematical model based on the van der Pol equation, which has been developed by Wever (1962, 1963, 1964) to describe circadian rhythms. That model places strong emphasis on the tonic (i.e., parametric) effects of light upon a circadian rhythm, as do Coupled Stochastic Systems; but

Wever's conceptual scheme makes no direct provision — as Coupled Stochastic Systems do — for the possibility that the tonic action of light might result in either acceleration or slowing of the pacemaker, depending upon phase within the rhythm at which the light occurs. Instead, Wever postulates, for a diurnal animal, that the tonic action of light would always accelerate the system, with only the amount of acceleration being dependent on time at which the light is given. Both models therefore predict that light of, say, 10 h duration should produce a shortening of the free-running period, as seen in the experimental data. The parametric component of Wever's model, however, predicts that the longer the duration of the light, the shorter the period of the rhythm: the converse of the results obtained in the experiment.

However, as Wever has pointed out to me (pers. comm.), the experimental results of Figure 14.1 can be readily reconciled with his model, by invoking the phasic (i.e., differential) effects of the light-out stimuli. In terms of Wever's model, therefore, these data would simply indicate that the phasic effects of light can override those effects expected due to the tonic influence of supplementary lighting. This experiment thus did not prove satisfactory as an attempt to reduce uncertainty by eliminating alternative hypotheses (cf. Introduction of Chap. 13). It simply represents one more kind of experimental data which can be adequately accounted for, at least qualitatively, by a Coupled Stochastic System.

Nevertheless, it appears to me that clamped free-running experiments have significant potential for elucidating the effects of light upon circadian systems. Most important is the manner, described above, in which they can be used to evaluate the influence on the animal of that light intensity which prevails at the time of activity onset, which is of key relevance to a Coupled Stochastic System. In addition, the possible phasic effects of the light-off stimulus, which complicate the interpretation of the experiment shown in Figure 14.1 in terms of Wever's model, could be evaluated, if the light stimuli of different duration administered in such a clamped free-run experiment were not square-wave pulses, but instead incorporated a gradual decrease in intensity at the end of the stimulus.

14.2 Two Problem Cases

The following sections describe two qualitative expectations which follow from Coupled Stochastic Systems and which are demonstrably wrong. Deficiencies of the models are thereby exposed, and we will consider ways in which those problems might be remedied, and the physiological implications of such modifications.

14.2.1 The Relationship Between the Discharge Coefficient and Excitation

The first of these problem cases is associated with conclusions drawn in preceding chapters. In determining the mechanism responsible for phase advance of a Coupled Stochastic System, with nocturnal-animal formulation of threshold, it became evident (Chap. 10) that the requirement was for unentrained pacers which, if prevented from discharging at the end of a given bout of discriminator activity, have the capacity to affect subsequent onset of activity. This demanded relatively large values of $\bar{\delta}$, the Discharge Coefficient, and of ϵ, the Excitation produced by the discriminator; values of 0.5 and 12 h proved satisfactory. In Chap. 12, it became evident that long-persistent after-effects of a given stimulus regime were dependent on the capacity of an unentrained pacer to become invisible for many successive cycles, during which it would not contribute to the next onset of discriminator activity. This kind of performance could apparently be achieved only by small values of $\bar{\delta}$ and ϵ; most calculations in that chapter were based upon values of 0.2 and 3 h, respectively.

These two aspects of system performance — phase advance by light pulses late during the animal's wakefulness, and long-duration after-effects — thus would seem to be mutually incompatible within the framework of a Coupled Stochastic System, and the prediction follows naturally that no nocturnal animal should show both kinds of behavior. Data for several rodents clearly contradict that expectation (Pittendrigh and Daan 1976a and Daan and Pittendrigh 1976a). Let us consider the extent to which the models must be modified, in order to accommodate the two sorts of apparently irreconcilable requirements.

A remedy can be achieved by the simple assumption that the ensemble of pacers includes some elements which contribute to fulfilling one set of requirements (large $\bar{\delta}$ and ϵ) and others that fulfill the second requirement (small $\bar{\delta}$ and ϵ). With the parameter γ, the pacers of the ensemble have already been given a spectrum of values for duration of discharge, but the effect of feedback from the discriminator in all previous simulations (except those which led to parts E, F and G of Figs. 6.11 and 6.12) has been represented by a single parameter ϵ. It is now required that ϵ be quite small for some pacers (say 3 h) and quite large for others (say 12 h). Furthermore, it would be extremely convenient if ϵ_i, the effect of feedback on the i th pacer, be correlated with duration of discharge of that pacer, so that pacers with short discharge tend to have small values of ϵ. (This correlation is not, in fact, absolutely demanded by the experimental data, but would serve to enhance the magnitude of both after-effects and of phase advance, by assuring that all unentrained pacers with short discharge could contribute to after-effects; and that all with long discharge could potentially contribute to phase advance.) Both the existence of inter-pacer variability in ϵ, which is essential, and the desired correlation between ϵ and the duration of discharge can be conveniently specified by invoking only a single additional parameter; for example, it could be stipulated that $\epsilon_i = \overline{\epsilon} + \rho \, (\delta_i - \overline{\delta})$, with ρ, the coefficient of proportionality, being the new parameter. [We must now also incorporate a subscript for ϵ in Eq. (6) of Chap. 4, so that $\overline{X}_{a,i} = \overline{X}_{b,i} - \epsilon_i$.]

The physiological implications of this proposed correlation are noteworthy. A small value for ϵ means that the intrinsic cycle length, the basic period of the pacer, will be very little affected by feedback; a large value for ϵ means greater feedback-induced increase in pacer frequency. Hence, by analogy with ordinary neuronal behavior, the magnitude of ϵ can be considered to be a reflection of how "sensitive" the pacer is to the sort of "standard stimulus" administered via the discriminator. The Discharge Coefficient, δ, can also be interpreted in a qualitative way as a reflection of a pacer's readiness to respond, its excitability. A pacer with a small value for δ requires a longer recovery interval in order to produce discharge of a given duration than one with a large value for δ. The correlation invoked above suggests that these two components of pacer excitability may reflect some common physiological process within, or property of the cell.

Whatever the physiological interpretation, however, it should be recognized that this kind of relationship involves a significant elaboration of a Coupled Stochastic System, an admission that the models in their elementary form are seriously incomplete. It turns out, in fact, that the conceptual elaboration required goes somewhat further than indicated here, because of the additional requirement that elements with small values of ϵ_i and δ_i, as well as elements with large values of ϵ_i and δ_i, both be members of the unentrained subset of pacers. An exploration of the consequences of this requirement is offered in Appendix 14.A. Meeting these several requirement places severe constraints on the set of models which are satisfactory candidates for the pacemakers of those nocturnal rodents which show both long-duration after-effects and a capacity for phase advance of the rhythms due to light pulses. While the models can be restored to viability by such restrictions, they thereby become more complex, both by a conceptual elaboration and by the incorporation of the additional parameter, ρ.

14.2.2 Splitting of an Activity Rhythm

The second critical failure of Coupled Stochastic Systems involves what is known as "splitting" of a circadian rhythm. The problem arises as follows: due to the nature of the system envisioned and the coupling proposed in Eq. (6) [either Eq. (6) on p 38, or its modified form, suggested above], the usual expectation is that an animal under constant lighting conditions

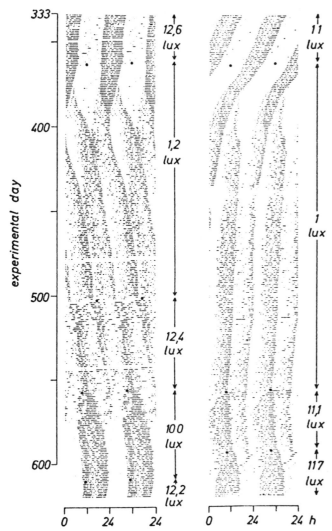

Fig. 14.2. Two examples of splitting of the circadian locomotor pattern of tree shrews (*Tupaia belangeri*) into two components, and rejoining into one component, due to change in light intensity. Note that each record is double-plotted, with each 24-h segment repeated, so as to simplify visualization of trends which scan a single day. The details of the activity pattern during the initial transition stage (days 375–400) are of particular interest. This section of the record indicates that the two components, which are gradually drifting apart, were initially of similar duration and had been completely overlapping in their phasing. From Hoffmann (1971)

will either have a circadian and clearly unimodal pattern of locomotor activity, or — under extremes of threshold (cf. Chap. 7) — become arrhythmic. Under peculiar sorts of entrainment regimes, bimodal patterns of activity might in principle be evoked, which could then persist, as a metastable state, under constant conditions (cf. Fig. 12.8). The elementary class of models considered here, however, predicts that a transition to such a bimodal pattern could not slowly develop over many cycles, while the animal is held under constant conditions.

This prediction is clearly contradicted by certain experimental data. Rhythm splitting under constant conditions has been thoroughly documented as a reproducible phenomenon in the activity pattern of the diurnal tree shrew, *Tupaia belangeri,* when the animals are exposed to quite dim levels of constant light (Hoffmann 1971; examples in Fig. 14.2). In these remarkable records, the experimental data seem to demand the existence of two separate, semiautonomous oscillatory systems, which are, under bright constant light, fully in synchrony with each other; and which can, under dim light, gradually drift out of phase with each other and then lock on again in approximate antiphase. Other instances of rhythm splitting, which tend to be less clearcut than those in the *Tupaia* but seem to be similar in principle, have been occasionally observed in less exotic species, including a few cases in the hamster and the deermouse, when exposed to continuous light (Pittendrigh 1967; Fig. 14.3). Here is, then, very clearcut failure of Coupled Stochastic Systems to account for experimental data, of a sort which is particularly informative.

The elaboration of the models necessary to accommodate such experimental results is more than superficial; instead of a single pacemaker system of the sort envisioned until now, it appears to me that two are required, or at least a pacemaker system with the equivalent of two discriminators, such as that illustrated in Figure 14.4. Note that interconnections between the ensembles of the pacers *via* the discriminators are essential. If the discriminators in such an interconnected array are assumed to have equivalent thresholds (expressed as proportions of the number of pacers which provide input) and to respond identically to changes in light intensity, then, as simulations have demonstrated, the summed output of the two discriminators appears to be entirely equivalent to that of a model with only one discriminator and somewhat fewer pacers. If, however, the thresholds of the two discriminators are assumed to respond somewhat differently to light intensity, then one can induce the rhythmic system to split into two components, one from each discriminator, by an appropriate choice of threshold values. Simulations have demonstrated that with a modest degree of interconnectedness, such as that shown in Figure 14.4, the two split components of the resulting rhythm would free-run with different periods; but that when the two components are again "in phase," they have a tendency to lock on rather than to free-run indefinitely. This kind of result is comparable, then, with experimental observations like the hamster recording of Figure 14.3.

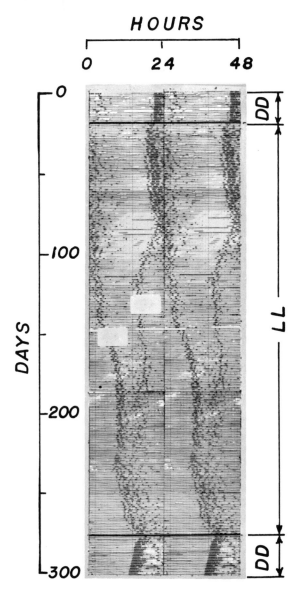

Fig. 14.3. Splitting in the activity pattern of a hamster (*Mesocricetus auratus*) into two components under bright light, followed by spontaneous refusion of the components into a unimodal pattern. Double-plotted record, with constant-light treatment (LL) preceded and followed by continuous darkness (DD). From Pittendrigh (1967)

If, instead of different responses of the two discriminators to light intensity, one postulates that there is weak antagonism between the two discriminators, a kind of inhibition such that each discriminator, when active, tends slightly to raise the threshold of the other, parameters can be assigned to the model so that activity of the two discriminators remains in phase when threshold is low. When threshold is raised to a higher level, the two components of the rhythm gradually move out of phase with each other, and finally lock on with a timing, such that there is no overlap in the activ-

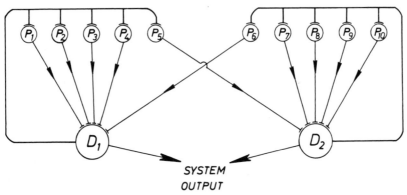

SYSTEM
OUTPUT

Fig. 14.4. Schematic diagram of an interconnected, paired pacemaker system, of the sort which is apparently required, within the framework of the models, to account for experimental observations of rhythm splitting under constant conditions. The two discriminators, D_1 and D_2, are interconnected by means of some fraction of the pacers (P_5 and P_6 in this diagram). These connections provide weak mutual excitation which would tend to keep the output of the two systems in phase with each other, unless other factors (see text) intervene

ity of the two discriminators: 120° to 180° out of phase, in the simulations that have been performed. Thereafter, the two rhythm components run parallel, with the same period, until threshold is again lowered, at which time they gradually return to the initial, in-phase relationship. Simulations of this kind of result therefore greatly resemble the kind of rhythm splitting seen in the *Tupaia* (Fig. 14.2).

Again, then, it is possible to rescue the conceptual scheme underlying Coupled Stochastic Systems from rejection by means of bold ad hoc hypotheses, which are, I think, compatible with the idea of pacemaker systems involving neuronal substrate. The conceptual elaboration required here, however, is extensive, both in terms of morphological implications (two discriminators, with cross connections) and in terms of the number of parameters required to characterize the complete system. Although a crucial weakness of the elementary models has been exposed here, it should be emphasized that the behavior of a model like that of Figure 14.4, when in the unsplit condition, is essentially indistinguishable from that of the elementary models. This means that the other predictions which were derived for simple, single-discriminator models (Chap. 13) are not invalidated or even in need of major revision. We need only specify that those predictions are not intended to be applicable to that rare situation in which an animal's rhythm has undergone splitting.

The physiological implications of an extended model like that of Figure 14.4 are worth noting. In Chap. 8, the possibility was mentioned that the role of the discriminator in a Coupled Stochastic System could in principle be subsumed by an excitatory neurohormone, secreted by each pacer during its discharge phase and affecting all other pacers of the ensemble. Severe constraints on that possibility would arise with a complex pacemaker like that of Figure 14.4. Such a model demands that each discriminator interact only with its own associated ensemble of pacers. It is difficult, therefore, to envision how the excitatory influence of the discriminators upon "their" pacers could be mediated by a single hormone which is present in the general circulatory system. Neurosecretion is not, of course, excluded as the vehicle by which coupling among such pacers might be achieved, but if that arrangement exists, the influence of the secreted substance must, I think, be severely confined spatially, in order to reconcile such a two-discriminator model with the split-rhythm data from the *Tupaia*.

The possible role of hormones in a Coupled Stochastic System will be further considered in Chap. 15. While the experimental data involved there are complex, the demands which those data place on the models are by no means as severe as those which arise from rhythm splitting.

Appendix 14.A. Reconciling Phase Advance with After-effects

In order to account for the occurrence of both after-effects and phase advance in the wake-sleep rhythms of nocturnal rodents, the models, in their elementary form, demand that some pacers in the ensemble have small values of δ_i and ϵ_i, and others have large values for these parameters. Beyond this, it is required that many of the pacers with small values of δ_i have basic periods shorter than that of the ensemble; and it is required that many of the pacers with large values of δ_i also be members of the unentrained subset, so that their discharge will quite often begin later during the hours of wakefulness. If it is presumed, as is the case with the type model, that unentrained pacers are only those with basic periods shorter than that of the pacemaker ensemble, then these requirements cannot be met by a model based on the equations of Chap. 4. The problem arises because of Eqs. (2) and (3), where the average duration of recovery is specified as $\overline{X}_{b,i} = \overline{X}_b + \beta z_{i,\beta}$; and duration of discharge as $Y_{i,j} = \delta_i X_{i,j}$. The consequence of these relationships is that pacers with large values of δ_i will generally have basic periods appreciably longer than those with small values of δ_i. With this kind of formulation, one cannot readily have significant numbers of both kinds of elements as members of the unentrained, short-period subset.

A reformulation that would remedy the problem is the following. Let the average basic period of the *i* th element, symbolized as $\overline{P}_{b,i}$, equal $\overline{P}_b + \beta z_{i,\beta}$, where $\overline{P}_b$ is the ensemble average of the value for basic period, and $z_{i,\beta}$ and β serve to specify inter-pacer variability in a manner similar to that of Eq. (2) of Chap. 4. For the average duration of the recovery phase, let $\overline{X}_{b,i} = \overline{P}_{b,i}/(1 + \delta_i)$, so that $X_{b,i,j} = \overline{P}_{b,i}/(1 + \delta_i) + \alpha\, z_j$ in the j th cycle, where $X_{b,i,j}$, δ_i, α, and z_j are as in Eqs. (1), (3), and (4) of Chap. 4; and let the duration of discharge, as before, be $Y_{i,j} = \delta_i\, X_{i,j}$. It should be noted that the introduction of $\overline{P}_b$ does not, in principle, add another free parameter to the models. The equations of Chap. 4 do not explicitly deal with $\overline{P}_b$, but its equivalent can be derived from parameters there, being equal to $\overline{X}_b\,(1 + \delta)$. In the present context, $\overline{P}_b$ serves as a scaling factor, as did $\overline{X}_b$ in the equations of Chap. 4; and $\overline{X}_b$ has become a derived quantity, determined by $\overline{P}_b$ and δ_i; it is not a free parameter, as in the previous formulation.

The physiological implications of this reformulation are worth brief consideration. The equations of Chap. 4 have the consequence that if a given pacer has a consistent tendency for long duration of discharge, it is also very apt to have a basic period which is unusually long. The equations proposed here serve to eliminate that correlation. The formulation of Chap. 4 implied that the circadian cycle of a pacer arises as the sum of two semi-autonomous processes, recovery and discharge, each with its own independent time course. Here, instead, we assume that the basic temporal unit of a pacer is its circadian period. Some pacers have only the capacity for a short discharge and others can do more, and that capacity has no bearing on whether the pacer is a fast or a slow member of the circadian population. In addition to mathematical reformulation, then, we have here introduced a modification in the mechanisms by which a pacer's circadian cycle is presumed to originate.

More fundamental modifications of Coupled Stochastic Systems could also, of course, remedy the problem addressed here. One might, for example, consider an alternative process for generating phase advance in the rhythm of a nocturnal rodent, perhaps invoking unentrained pacers with basic period very much longer than that of the ensemble, so that even with feedback they remain unentrained. Figure 14.5 presents in schematic form how such a pacer might be able to induce phase advance of the ensemble, due to light stimuli which prematurely terminate feedback.

Preliminary attempts to simulate this process on the computer indicate further constraints on the model. The onsets of discharge by those long-period pacers, which should permit phase advance of the pacemaker by the process shown in Figure 14.5, will be more or less evenly distributed throughout the entire duration of feedback, without the peak illustrated in Figure 10.6. Appreciable phase advance therefore requires a great many unentrained, long-period pacers. To reconcile this requirement with the demand that many unentrained, short-period pacers be available to produce after-effects, with periods very near to that of the ensemble, apparently requires that the values of δ_i be selected from some parent population which differs markedly

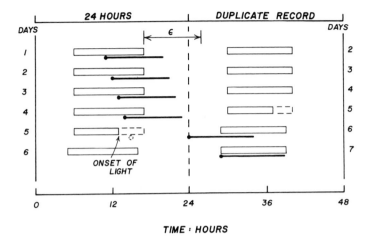

Fig. 14.5. Schematic illustration of how phase advance of the pacemaker might be mediated by pacers with very long period. At the onset of light, feedback is terminated, so that the pacer, which would otherwise have begun discharge at the time indicated by the *dashed circle,* is postponed by ϵ hours, thereby permitting its discharge to contribute to next onset of feedback. The requirements for this process are that $\epsilon + Y_b$ be greater than the duration of system inactivity (see Fig. 10.5), and that the basic period of the pacer, less ϵ, be greater than the period of the ensemble rhythm. *Open rectangles* correspond to system feedback; *heavy horizontal lines* represent discharge by the pacer

from the Gaussian distribution: perhaps a bimodal distribution or perhaps a strongly skewed distribution. I have not further explored this alternative, but the complications encountered suggest that the reformulation proposed earlier is probably the simpler way of remedying the problem addressed here.

Chapter 15
Morphology of the Models: Where is the Pacemaker?

In formulating the behavior of the individual pacer elements of the pacemaker ensemble, several concepts were incorporated which have their origin in the behavior of high-frequency neurons. The recovery phase of the pacer, with its associated stochastic variability, is in some senses related to the idea of neuronal "refractory time", and even more closely related to the "recovery" process which is envisioned in several formal models proposed for the behavior of spontaneously firing neurons in the high-frequency domain (cf. Figs. 4.1 and 4.2). Furthermore, the idea that excitatory discriminator input does not affect the rate of the pacer recovery process, but only alters the excitatory state associated with onset of discharge (cf. Figs. 4.4, 4.5, and 4.6) is also closely related to concepts of formal models which have been developed to describe the influence of tonic stimuli on high-frequency neurons. Now that we have explored in some detail the performance of a particular sort of hypothetical network consisting of such elements, the question naturally arises, "To what extent can these models be extrapolated back into a neurophysiological context, as guidelines for the experimentalist in his search for the biological clock? " More explicitly, what clues do the models provide which might facilitate the identification of structures within the organism which could correspond directly to elements of the hypothetical pacemaker? In consideration of this issue, the pacers and the discriminator will be dealt with separately.

15.1 The Pacers

One of the most essential properties attributed to the pacer ensemble is that each of the elements of the array be capable, at least in principle, of independent oscillations with a period which is more or less circadian. The idea of independence is central here; *absolutely* independent behavior in the intact animal is not demanded, but at the least an unentrained pacer must be able for many cycles to go its own circadian way, without being significantly influenced by its phase relationship to the behavior of the ensemble or that of the whole animal. The kind of autonomous behavior envisioned here suggests that each pacer of the model should correspond to

some sort of discrete morphological entity, separated from other similar
entities by boundaries which restrict interactions. Since the behavior of the
pacer has, in addition, been specified in a way which has similarities with
that of high-frequency neurons, consider for the moment the proposition
that the pacers are single neurons, and that their circadian activity is moni-
torable by spike activity. How might the search for such neurons be undertaken?

In the intact animal, showing normal circadian rhythms, it is to be expected that the vast majority of cells in the brain will show circadian variation in their output. Simple demonstration of circadian variation in the behavior of a given kind of neuron there is thus not apt to be particularly informative. If it were possible to remove a selected small portion of the brain,
maintain it in culture, and demonstrate persistence in the tissue of circadian
rhythmicity, then that brain area would be a plausible candidate as the site
of the pacemaker envisioned here. This sort of search technique, however,
would depend heavily upon finding adequate culture conditions; the prob-
able failure of such attempts would of course not permit any conclusions
about the presence or absence of the pacemaker.

A more promising approach is to take advantage of certain unusual circumstances in which the whole-animal rhythm, the wake-sleep, activity-rest
cycle, is suppressed, although subsequent observations demonstrate that
the pacemaker had been functioning normally in the interim. Two different
opportunities of that sort, which have apparently not yet been exploited,
suggest themselves. As shown in Figure 3.1, certain drugs can suppress wakefulness on a short-term basis, without producing significant reset of the pacemaker. In an animal subjected to such treatment, one should expect to find,
in the pacemaker itself, an unperturbed circadian cycle, comparable in its
phasing and amplitude with that when no drug is administered. Electrodes
implanted in those brain areas suspected of harboring the pacers ought to
permit a clear distinction between those central-nervous-system circadian
rhythms which arise as a consequence of wakefulness itself, and those which
more directly reflect pacemaker behavior. Furthermore, since the phase of
the pacemaker cycle which is ordinarily associated with onset of wakefulness in the models is characterized by the explosive "reset" of many pacers,
one can then see whether there is a similar "explosive" event in the postulated pacer ensemble at the *expected* time of onset of wakefulness, even
when wakefulness is suppressed by the drug.

Another opportunity for this sort of approach is found in the behavior
of short-time hibernators. French (1977), for example, has shown that under
appropriate temperature conditions, the pocket mouse *Perognathus longimembris* can be brought into a weak hibernatory state in which the animal
awakens regularly every second day at times clearly dictated by an endogenous circadian pacemaker. Similar data from this species have been also

published by Lindberg et al. (1971). This situation appears to be an excellent one in which to search, with implanted electrodes, for brain areas showing relatively normal circadian cycles, whether the animal in fact awakens on that day or not. An autonomous pacemaker of the sort envisioned here ought to show such cycles.

An important clue from the models, which should assist in unequivocal identification of the pacer ensemble, is suggested by Figure 5.4. As a consequence of the nature of the coupling envisioned here, many of the pacers of the ensemble remain unentrained, and those unentrained pacers are characterized by a period shorter than that of the ensemble itself; pacers with average period somewhat longer than that of the ensemble are rare if not totally absent. If it were to prove feasible then, to follow, by means of implanted electrodes, the behavior of individual neurons — suspected of being circadian pacers — over several days, the models suggest that a significant fraction of the cells should have an average circadian period which is somewhat shorter than the whole-animal rhythm; average single-cell periods somewhat longer than the whole-animal rhythm, however, should be rare or absent within the pacer ensemble.

This behavior represents a diagnostic test for the recognition of any pacemaker ensemble which operates, even approximately, by the coupling principles envisioned in the present models. Any ensemble of circadian oscillators in which all elements are mutually synchronized, with an average period identical to that of the whole animal, or in which elements exist with average period only slightly longer than that of the ensemble, must operate by a form of coupling radically different from that envisioned here for the pacers. Should it turn out that the circadian pacemakers of birds and mammals do *not* include many elements with circadian period somewhat shorter than the whole animal's rhythm, then the models developed here would have negligible congruence with reality; they would describe a way in which circadian pacemakers *might* operate — but in fact do not.

15.2 The Discriminator

In the search for a structure or phenomenon which fulfills the role of the discriminator in the pacemaker models, very few direct guidelines are available. The computer studies described here offer no assurance that a discriminator ought even to exist as an identifiable entity. As indicated in Chap. 8, its function might in principle be subsumed by an appropriate nonlinearity in the manner in which each pacer of an ensemble *directly* interacts with all the others. The effect of light upon the pacemaker ensemble has been envisioned as mediated by shifts in the threshold of the discriminator, but this function as well could, in principle, be accomplished by a light-induced modification in the direct coupling interaction among pacers of an ensemble.

Even assuming, for the sake of discussion, that the discriminator may have a distinct functional counterpart, there is no assurance that it provides excitatory input to the pacers. As discussed previously (Chap. 8), the formal equivalent of excitation might be provided by a discriminator which inhibits pacer discharge, and which then has phasing of its "activity" opposite to that presumed here on the basis of excitatory coupling. For the sake of concreteness, Figure 4.3 illustrates the discriminator as a single "master" neuron, but there is no clear basis for assuming — as suggested above for the pacers — that the role of the discriminator ought to be fulfilled by nervous tissue, rather than, say, by a secretory organ. Beyond this, there is good reason for suspecting that Figure 4.3 is oversimplified is illustrating the discriminator as a single cell. We have seen that noise in the responsiveness of the discriminator potentially can limit system precision (Chaps. 6 and 8); a unicellular discriminator might be expected to exhibit more stochastic variation than would be shown by the average performance of a multicellular complex.

The only clear expectations regarding the identity of the discriminator are (i) that its activity — if it is a separate entity — be unequivocally in phase (or antiphase) with average cycles of the pacer ensemble, (ii) that its output affect the timing of the pacers, (iii) that its output arise as a strongly nonlinear function of summed pacer discharge, and (iv) that its output be modifiable by light intensity. By these criteria, it would appear that identification of the discriminator demands prior identification of the pacers, but in fact the situation is by no means as discouraging as these considerations seem to imply. Certain experimental data suggest the exciting possibility that a discriminator, or at least a portion of one, may have already been discovered, although not identified as such.

The experiments of interest in this context were begun in the laboratories of Prof. Michael Menaker, and center around the role of the pineal organ in circadian rhythmicity of birds, particularly the house sparrow. A brief summary of these experimental findings is as follows:

1. Surgical blinding of the sparrow does not abolish the responses of their circadian systems to light (Menaker 1968), and similar results have been obtained with several other species of birds. The rhythms of blinded birds can be entrained by light cycles of very low intensity, mediated by light receptors located somewhere in the brain, which have not, to date, been further identified or localized.

2. Circadian entrainment of sparrows by light cycles is also possible after removal of the pineal organ (which, at least in lizards, is known to be photoreceptive); entrainment by light is also possible when both the eyes and the pineal are removed (Menaker 1971), showing that the extraretinal light receptor involved in circadian entrainment is not — or is not *just* — the pineal.

3. House sparrows, whether blinded or not, show clearcut persistent circadian rhythms in darkness; but following complete pinealectomy the birds have not been observed to

sustain a circadian rhythm in darkness (or, for that matter, in constant dim light). Control surgery does not have this effect (Gaston and Menaker 1968).

4. Following incomplete pinealectomy of the sparrow, however, there is a poor correlation between the amount of pineal tissue removed and persistence or loss of rhythmicity; even in birds with only scattered parenchymal cells of presumed pineal origin, sustained rhythms have occasionally been observed (Gaston 1971).

5. There are important interspecies differences among birds in the significance of the pineal organ to circadian rhythmicity. Results like those in points 3 and 4 have also been obtained with the white-throated sparrow (McMillan 1972), the white-crowned sparrow (Gaston 1971), and the house finch (J. Fuchs, pers. comm.); but removal of the pineal organ does *not* ordinarily lead to loss of sustained circadian rhythms in the European starling. Gwinner (1978) found that only 3 of 26 starlings became permanently arrhythmic following pinealectomy. (See also Rutledge and Angle 1977.)[15] Nevertheless, the pineal organ is not irrelevant to rhythmicity in the starling; pinealectomy induces a shortening of the circadian period, and leads to large perturbations of the activity pattern, including decreased precision of the rhythm and a more diffuse temporal distribution of activity.

6. Pinealectomy of the house sparrow does not necessarily destroy circadian rhythmicity immediately, but instead only destroys the capacity of the birds to *sustain* a rhythm for long periods under constant conditions. If the pineal is removed from a sparrow during entrainment by a light-dark regime, and the bird is transferred to darkness following surgery, the circadian rhythm very gradually decays, with damping of the rhythm extending over intervals of from a few days to 2 weeks (Menaker 1971; Menaker and Zimmerman 1976).

7. When a pinealectomized house sparrow is subjected to a light-dark cycle, there is clear synchronization of the animal's rhythm (point 2, above); moreover, this is not simply passive driving of the wake-sleep cycle by light, but typical entrainment of an endogenous rhythm, with daily onset of activity often many hours before onset of the lights. Such an entrained, pinealectomized sparrow, when transferred to darkness, then shows a clear free-running circadian rhythm which gradually deteriorates over the next several days, with arrhythmicity ensuing (as in point 6) within 2 weeks or less. Rhythmicity can be temporarily restored again and again by entrainment with light, and even a single light pulse can initiate a circadian rhythm which persists for a few days in darkness (Menaker 1971; Menaker and Zimmerman 1976).

8. When a sparrow is entrained by a light-dark cycle and then pinealectomized, its phase relationship to the light-dark regime changes systematically, in a way which suggests that the free-running period of some rhythmic center has been shortened relative to that in the intact bird. For example, if the intact bird began its daily activity 5 h before onset of lighting, it will, following pinealectomy, begin its activity earlier − perhaps 8 h before onset of light, perhaps even earlier (Gaston 1971).

9. When the pineal of the sparrow is deprived surgically of its neural output, but left in place, the bird will usually maintain the ability to sustain rhythms in darkness; chemical intervention, intended to destroy in addition the neural input to the pineal organ, was also without noticeable effect on persistence of circadian activity rhythms (Zimmerman and Menaker 1975).

10. When a pinealectomized house sparrow, which is arrhythmic in darkness, receives a pineal organ implanted into the anterior chamber of the eye, it may (in ca. 40% of the cases) recover the ability to show sustained rhythms in darkness, provided that the pineal tissue is sliced before implant (Zimmerman and Menaker 1975).

---

15 Of the Japanese quail, Simpson and Follett (1979) report that pinealectomy has no detectable effect upon circadian rhythmicity.

11. Often, the recovery of rhythmicity described in point 10 arises within a day or two of receipt of the donor tissue, and in such cases, the light history (prior entrainment regime) of the donor bird has a detectable influence upon phasing of the circadian rhythm of the recipient (Menaker and Zimmerman 1976; Zimmerman 1976).

12. Melatonin is one of the primary secretory products of the pineal organ. In the pinealectomized European starling (which — see point 5, above — retains its circadian rhythm), it is possible to synchronize the bird's circadian rhythm by daily injections of melatonin. Control injections without melatonin do not usually have this effect (Gwinner and Benzinger 1978).

13. Implants of capsules filled with melatonin into intact house sparrows either produce arrhythmicity or, if not, induce some shortening of the period of the bird's persistent rhythm (Turek et al. 1976).

This series of observations clearly shows that the pineal organ, and its product, melatonin, are somehow often involved in the pacemakers of birds and in their synchronization, but the results are surprisingly complex and some of the findings from different species contradict each other. It is evident, of course, even in the sparrow, that the pineal is not the *entire* pacemaker (points 6 and 7), but it is not at all clear what the nature of its involvement in circadian timing might be.

Two related interpretations of the data have been proposed. Menaker and Zimmerman (1976) suggest (for the sparrow) that the pineal organ is a "master oscillator" which has an endogenous circadian period and which is capable of sustaining its own circadian rhythm indifinitely; that the pineal drives another oscillatory system, located elsewhere in the brain, which is capable only of a damped circadian rhythm for a few cycles; and that both the pineal and that other oscillatory system are independently responsive to light. Gwinner (1978) also postulates that the pineal is a master, driving oscillator with the capacity for independent circadian rhythms, with input to an oscillatory system located elsewhere; but suggests that this other, driven oscillatory system consists of an array of loosely coupled oscillators, each of which, at least in principle, could produce a sustained circadian rhythm. Gwinner then suggests that if the coupling within that array is sufficiently weak, loss of phased pineal input would result in loss of mutual phase synchrony among the oscillators of the array; progressive gradual damping of the animal's rhythm would then ensue (sparrow). With stronger mutual coupling within the array, loss of sustained pineal input would still permit persistence of a (degraded) rhythm in the animal, as seen in the starling.

Careful consideration of these results suggests to me that they might also be satisfactorily interpreted by the hypothesis that the pineal organ constitutes a *part* of the discriminator of the pacemaker models under consideration here. That is, it contributes to the coupling among pacers (located elsewhere), but need not be a circadian oscillator, much less a driving master oscillator; it need not even be capable of self-sustained circadian rhythmicity.

One way that might work is the following: suppose that melatonin, when present in the circulatory system, exerts an inhibitory influence upon the circadian pacers. An absence of melatonin thereby becomes the formal equivalent of excitatory discriminator activity. Suppose, further, that pacer discharge, when of sufficient intensity, can directly or indirectly inhibit the synthesis and/or release of melatonin, thereby accelerating all nondischarging pacers; and that light directly inhibits the synthesis and/or secretion of melatonin to an intensity-dependent degree. It is clear that the pineal is the dominant source of melatonin in some species of birds (Pelham 1975), but suppose that there exist other sources of melatonin in the brains of the sparrow and the starling; and that all sources are similarly inhibited by light. The experimental observations described above — with only one exception — can be accommodated by this interpretation, as described below:

1. The pineal organ, as well as all other nonpineal sources of melatonin, together may constitute the extraretinal light receptor of birds (observations 1 and 2, above). The multiplicity of melatonin sources might account well for the fact that the extraretinal photoreceptor(s) of the sparrow have not yet been identified, 10 years after their existence was documented.

2. For a bird in which the pineal is an important source of melatonin, pinealectomy would greatly reduce the amplitude of feedback cycles (Fig. 15.1). That reduction would weaken the mutual coupling within the pacemaker ensemble; it would decrease the value

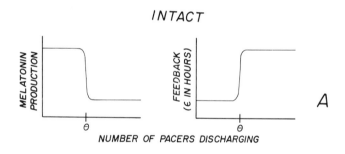

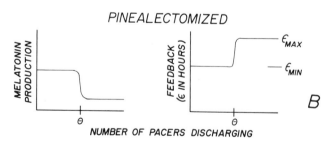

Fig. 15.1 A and B. Schematic diagram showing the postulated relationship between melatonin and excitatory feedback in the pacemaker models, in an intact bird (**A**) and a pinealectomized bird (**B**). Melatonin is presumed to have an inhibitory effect upon the pacer, lengthening the duration of its refractory interval. Absence of melatonin then translates into the application of excitatory feedback to the pacers

of Excitation (ϵ) in the models (the difference between ϵ_{max} and ϵ_{min} in Fig. 15.1), causing some of the longer-period pacers to become unentrainable. With sufficient reduction in ϵ, the circadian rhythm of the pacemaker could not persist, due to loss of synchrony in the ensemble (Observation 3).

3. The period of the pacemaker models is directly proportional to the average recovery interval of the pacer ensemble, in the absence of feedback by the other pacers. The role of feedback (ϵ) in the models is then to shorten the recovery interval, when a sufficient fraction of the pacers are discharging. Note that in Figure 15.1B the minimum value of feedback (ϵ_{min}) would be greater than in the intact bird. This would correspond to a shortening of the average recovery interval of the pacers and would mean that the average period of the pacemaker should be shortened by pinealectomy. Hence, the shortening of the free-running period in pinealectomized starlings, as well as the increase in phase lead of pinealectomized sparrows, relative to a light cycle, are consistent with the postulated decrease in production of melatonin, and the resulting increase in the minimal value of feedback (Observations 5 and 8).

4. Interindividual differences, as well as interspecies differences in the results of pinealectomy (Observations 4 and 5) may reflect differences in the relative contributions of nonpineal sources of melatonin.

5. The fact that pinealectomy of the sparrow has somewhat different consequences, depending upon whether the bird was previously entrained by light or not (Observation 6) indicates that the pacemaker of the entrained bird is somehow in a state different from that during free run. The nature of the experimental result (gradual damping of the rhythm, following entrainment, rather than immediate loss of rhythmicity) might within the framework of the models envisioned here be taken to indicate that the pacers of the ensemble are more tightly clustered in phase, more strongly coordinated in their timing, under light-dark cycles than in free run. Such a difference could arise because of exogenous, transient effects of sudden light transitions upon melatonin production (Fig. 15.2). (Note that this interpretation is quite similar to, but in fact different from, the postulated effects of light transitions which were invoked in Chap. 11, and Fig. 11.2.) A tighter clustering of pacer phases in the entrained state, i.e., stronger mutual entrainment, would mean that in spite of reduced amplitude of feedback cycles following pinealectomy, pacemaker rhythmicity could persist for several cycles before deterioration into arrhythmicity.

6. Entrainment of the sparrow, following pinealectomy (Observation 7) would be expected if there were light-mediated inhibition of melatonin production in nonpineal sources, leading to resynchronization of the pacers.

7. The findings, (i) that de-enervation of the pineal does not impair rhythmicity, (ii) that rhythmicity can be restored in pinealectomized sparrows by implantation of pineal tissue into the eye, and (iii) that pinealectomized starlings can be entrained by regular melatonin injections (Observations 9, 10, and 12) are consistent with the interpretation that blood-borne melatonin is the vehicle by which the pineal exerts its effect upon the hypothesized pacers. Furthermore, the nature of the entrainment which results from injections of melatonin into the starling (activity onset following injection by several hours) is consistent with the idea that the influence of melatonin on the circadian pacemaker is an inhibitory one, able to postpone onset of discharge by the pacers.

8. The finding that when the pineal is implanted into the eye of a pinealectomized sparrow, the phase of the resulting rhythm can be affected by the light regime which the donor experienced (Observation 11), has been taken as very strong evidence for the idea that the pineal is the master oscillator, which ordinarily drives the circadian rhythm (Menaker and Zimmerman 1976). A sufficient explanation for the observation, however, is implicit in Figure 15.2. The implication of that diagram is that a sudden transition in light intensity produces a transient effect on the capacity of pineal tissue to release melatonin, an effect which decays over a matter of a few hours following transition in light intensity. Hence, pineal organs removed from birds at different times during a light

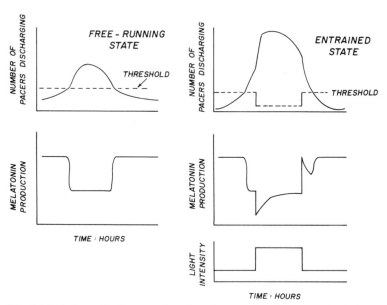

Fig. 15.2. Schematic diagram showing the postulated temporal relationship between melatonin production in an intact bird, and the state of the pacemaker, expressed in terms of the number of pacers discharging. The free-running state indicates the consequences of the relationship shown in Figure 15.1A; the entrained state shows the postulated additional consequences of an entraining light stimulus

cycle would be expected to have different impacts upon the pacer ensembles of recipients. The experimental observation demands that the pineal organ have some "memory" of the light regime to which the donor bird was exposed, but it could well be a relatively short-duration memory, without implying that the pineal is intrinsically rhythmic.

9. The results of chronic implantation of a capsule which releases melatonin, in an intact sparrow, cannot be accounted for by my interpretation, without some supplementary hypothesis. The interpretation proposed here in simplest form would predict that melatonin implants should lead to a lengthening of the bird's circadian period, because of an increase in the recovery interval of the pacers (Fig. 13.5), and not a shortening, as was observed. It is conceivable that secondary effects of exogenous melatonin may alter the secretion by normal sources of melatonin, as suggested by Turek et al. (1976), as well as by Gwinner and Benzinger (1978), but in the absence of evidence for or against that interpretation, this experimental result remains inexplicable by the present interpretation.

10. The preceding interpretation deals with how the pineal organ might affect the pacers envisioned by the models. The question of how the pacers might in turn affect the pineal organ — which is demanded by the models, if the pineal were to correspond to part of the discriminator — has not been explicitly addressed. The experimental evidence, while not yet compelling, strongly suggests that these effects, if they exist, might well be hormonal. Possible reservations to that interpretation are that the attempted chemical de-enervation of the pineal in the sparrow (Zimmerman and Menaker 1975) may have been incomplete; and that a pinealectomized sparrow, which has been restored to circadian rhythmicity by pineal tissue implanted in the eye, may have an abnormal and more diffuse rhythm than an intact bird. Discounting such reservations for the moment, and supposing that the effects of the hypothesized pacers upon the sparrow pineal are indeed mediated by hormones, one is led to the suspicion that the summed

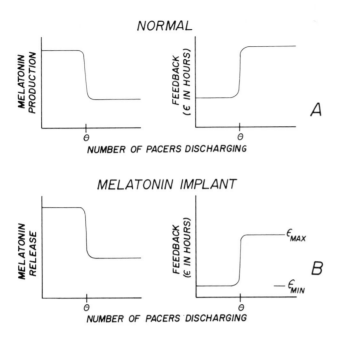

Fig. 15.3. Schematic diagram showing the relationship between melatonin and excitatory feedback expected, by the interpretation proposed in the text, during continual release of exogenous melatonin. The increase in melatonin, when few pacers are discharging, would represent a decrease in the minimum value of excitatory feedback, which would translate into an increase in recovery interval of the pacers and increase in period of the pacemaker

discharge activity envisioned here for the pacemaker models might be detectable, directly or indirectly, by the level of some hormone which is present in the general circulatory system. This would not, of course, require that the pacer ensemble itself be a secretory organ, but only that its output be transmitted to the discriminator via hormones.

The suggestions here about how the results of an elegant series of experiments might be accounted for within the context of the proposed pacemaker models involves the central idea that the pineal organ may function primarily in the coupling of circadian oscillators which are located elsewhere. The possibilities that a coupled-oscillator system may underlie circadian rhythms, and that pinealectomy in some way alters the nature or strength of the mutual coupling, are qualitative ideas which have been proposed by others, without invoking an extensive formal model of the sort developed in preceding chapters. For example, Underwood (1977), while accepting "strong evidence that the avian pineal is the site of a master driving oscillator", suggested that the pineal organ in *lizards* might well be a "coupling device between circadian oscillators in a multi-oscillator system." Gwinner (1978) also, while endorsing a master-oscillator role for the avian pineal, has suggested that interspecies differences in the results of pinealectomy in birds may reflect differences in the strength of coupling in a multi-oscillator system. The merit of the present models, however, is that they explicitly define the nature of the coupling postulated and thereby lead to a somewhat different view of the possible role of the pineal in that process. As a concequence, many kinds of testable prediction can be derived from the interpretation, which do not necessarily follow from the idea in its elementary qualitative form. Some of the more obvious predictions are listed below.

1. A pinealectomized sparrow is, by hypothesis, deprived of a major component of its extraretinal photoreceptor. More intense light should therefore be required for entrainment of such a bird than for an intact one; and because pinealectomy is postulated to reduce the amplitude of circadian variation in feedback (cf. Fig. 15.1), the range of entrainment — i.e., the range of period values permitting synchronization by a light cycle of a given intensity — should be reduced. These differences between an intact and a pinealectomized bird should be most pronounced in light cycles with gradual changes of intensity (cf. sinusoidal light cycles, Chap. 9), where the expected effects of sudden light transitions (Fig. 15.2) would be absent.

2. Arrhythmicity seen in birds kept in constant bright light would be here interpreted as due to suppression of melatonin production by light. It should therefore be possible to induce and entrain a circadian rhythm by means of melatonin injections, in a bird kept in light bright enough to produce arrhythmicity, just as Gwinner and Benzinger (1978) have done with pinealectomized starlings.

3. In the pacemaker models, it is unnecessary to presume that the discriminator has any intrinsic, self-sustained rhythmicity. Except for the postulated transient effects of light tansitions, the discriminator-mediated coupling of the pacer ensemble is viewed as a fully passive process. The interpretation proposed here for the role of the pineal organ therefore suggests that sparrow pineal tissue, maintained in vitro, might reach a non-oscillatory, time-independent state within a few hours of removal[16] thereby losing any capacity to transfer phase from the rhythm of the donor.

---

16 In contrast with this prediction, the interpretation of Menaker and Zimmerman (1976) for the role of the pineal organ in the circadian rhythms of birds leads to the expectation that isolated pineal should have the endogenous potential to show an undamped, self-sustained circadian rhythm under constant conditions. Several recent studies of chick pineal bear on this distinction.
When chicks have been held in light-dark cycles, and their pineal organs are then removed and held in vitro, that tissue releases an enzyme (serotonin N-acetyltransferase) which is involved in the synthesis of melatonin. Binkley et al (1978) observed a single pulse in release of this enzyme during organ culture; the timing of that pulse (from zero to about 15 h after removal from the chick), and even its very existence depended upon the prior lighting experience of the chicks. No persuasive evidence for a reproducible second peak in enzyme production was found during longer-duration tissue culture. Deguchi (1979) also observed a single large pulse of enzyme activity from isolated chick pineal glands. This pulse was synchronized by the prior lighting conditions experienced by the chicks, and was obtained when the cultured glands were held under constant darkness, but not under constant light. Deguchi (1979) also clearly demonstrated that chick pineal is directly responsive to light. When isolated pineals were subjected to a 24-h light-dark regime, enzyme activity continued to vary in phase with the lighting schedule for three days. In agreement with this latter result, Kasal et al. (1979) also found that when isolated chick pineals were held in a light-dark regime, enzyme activity was synchronized with the lighting conditions after three days in culture. In contrast with Binkley et al. (1978) and with Deguchi (1979), however, Kasal et al. (1979) observed a weak *second* peak in enzyme release from pineal glands cultured under constant darkness. That damped second peak occurred some 20 or so hours after the initial large peak, which was also reported by the the other research groups.

4. If the phasing of a recipient's rhythm by that of its pineal donor is mediated by the direct action of light on the donor, as hypothesized above, then the pineal of a donor with a free-running circadian rhythm should be incapable of imparting different phases to the recipient's rhythm. The phase of the recipient's rhythm should depend only upon when the implant is received, not on when it was removed from the donor.

5. The interpretation suggested here postulates that the entrainment of a bird by light cycles produces close phase synchrony of the pacers; and that the pacers of an arrhythmic pinealectomized bird have a diffuse distribution of phases, which can be reconsolidated by a pineal implant. The pacemaker of an entrained bird ought therefore to be less labile, less subject, if at all, to resynchronization by a pineal implant. Hence, cross-transplanting of pineal organs between two sparrows, which had previously experienced $180°$-opposite phasing of their entraining light cycles, should not, by the interpretation proposed here, lead the recipient birds to be seriously affected by the donor's phasing, during the post-transplant free run.

15.3 The Translation of Formalism into Concrete Morphology

The models developed in this treatise propose a particular form of coupling by means of which an array of circadian oscillators might interact in order to account for many interesting aspects of the whole-animal circadian behavior. The general purpose of this chapter has been to demonstrate that these models have potential beyond that of simply formally describing the

---

The results of Kasal et al. (1979), involving a second peak in enzyme activity under constant conditions, demonstrate that chick pineal has at least a weak capacity for endogenous rhythmicity, but none of the results obtained to date provides strong support for the interpretation that this rhythm is, or could be, self-sustaining. The strong damping of the rhythm shown in all three studies (one peak in Binkley et al. 1978 and in Deguchi 1979; two in Kasal et al. 1979) may, however, as those authors suggest, simply reflect deficiencies of the culture conditions. Eventually an undamped rhythm may be demonstrated in pineal tissue held under constant conditions.

On the other hand, the available results from tissue culture of pineal glands are entirely consistent with the idea that this organ may function primarily as a light-sensitive component of the hormonal coupling device (the discriminator), which maintains synchrony among self-sustained circadian oscillators located elsewhere (the pacers). In that case, long-term persistence of a circadian rhythm in isolated pineal tissue would not be demonstrable under constant culture conditions.

The interpretation suggested here requires only that the isolated pineal tissue of the sparrow vary in its output for a significant portion of a single cycle; and we now know (Kasal et al. 1979) that chick pineal can do somewhat more. If this capacity for strongly damped (two-cycle) circadian oscillations were to be incorporated into the models as a property of the discriminator, they would become more complex, although not fundamentally different in nature or behavior. There remains, however, the question of whether results obtained with the chick can be generalized to other birds, including the sparrow. This issue is particularly troublesome, in view of the previously mentioned differences in the effects of pinealectomy on activity rhythms among different species of birds.

whole-animal output and making predictions for novel sorts of experiments which also involve only the whole-animal output. Key characteristics of the hypothetical elements of which the model was constructed can also serve as guidelines for the experimentalist, as criteria by which he might recognize real morphological counterparts of the elements.

The detailed consideration of the experimental data on the pineal organ of birds serves as an example of how models of the sort envisioned here can potentially be translated into specific physiological contexts. This particular translation requires recognition that the discriminator of the models might be a secretory structure, rather than a neuronal one, and that the excitatory coupling envisioned could in fact be an absence of inhibitory coupling; but the basic formalism of the models remains unaltered by the translation. The example illustrates the fact that once a suspected counterpart of the model has been identified, further predictions are derivable, by means of which the postulated correspondence between model and animal can be further tested.

Chapter 16
A Reprise and Synopsis: On the Advantages of Apparent Redundancy

One of the most remarkable properties of the circadian wake-sleep rhythm of a higher vertebrate is its temporal precision. The cycle-to-cycle variability in period is often far less than 1% of the total interval; a bird or a rodent, kept under constant conditions without external time cues, may awaken day after day at times which are predictable to within 2 to 3 min. Speculation about the physiological mechanism which underlies such astonishing precision has led to the research described in this treatise.

It is conceivable, of course, that some very special sort of regulatory process might have arisen through evolution, to convey to certain single cells the capacity to generate extraordinarily reproducible circadian cycles. The alternative explored here is that the animal's clockworks may consist of an ensemble of many cells which are only more or less circadian in their performance: semiautonomous oscillators which are themselves not particularly reliable. The idea is that the extreme temporal precision evident in the behavior of the whole animal might arise through interactions within a network of erratic, unreliable components which are qualitatively similar to each other but by no means identical. In such an ensemble the many elements might appear to be redundant, but might actually serve as a means of developing order out of relative chaos.

One kind of elementary interactive network which might be considered as a candidate for this job is a system comparable with the vertebrate heart. There, a large array of interconnected pacemaker elements exists; the first of these elements which reaches a critical stage triggers all other units in the ensemble, and the process then begins anew. Elementary considerations demonstrate that this kind of network is not particularly well suited to the generation of precise system output. If the various pacemakers have different intrinsic frequencies, the discharges in such a system will be determined primarily by that pacemaker which has highest frequency. The system would be no more reliable than the single elements of which it is composed. Even if one assumes that the various pacemakers in the ensemble are fully identical in intrinsic properties, the expected precision would not be much improved. It can be shown that the temporal variability in synchronized discharge from an ensemble of 25 such units would be about half that of a single element, but beyond that stage, it requires nearly a million units to improve precision by another factor of two.

As a simple elaboration on that scheme, it is conceivable that the resetting or triggering of such an ensemble does not depend upon the first element to discharge, but instead requires that several of the independent, pacemaker-like elements have spontaneously reached their critical stage. In principle, it might be possible for the ensemble to perform the equivalent of counting: to wait until some predetermined number of events have occurred, at which time the balance of the system is triggered. An alternative which makes more sense in terms of physiological substrate is to postulate that each element which spontaneously reaches its critical stage continues thereafter for some time to give a signal, which says, in essence, "I've gotten there; come with me." The sum of these signals, from all supercritical elements, might then constitute the stimulus which can, when strong enough, trigger other elements in the ensemble to respond.

This scheme is, in its bare essentials, the kind of pacemaker system envisioned here. Within the postulated ensemble, each individual element, called a pacer, is presumed to have a quiescent or recovery stage, which is eventually terminated by randomly occurring (stochastic) events, and is followed by an "on" or discharging phase. When a sufficient number of these pacers are simultaneously discharging, their output constitutes a sufficient signal to affect the balance of the ensemble. The equivalent of a chain chain reaction would then follow, due to the achievement of a "critical mass": an explosive process by which the recovery intervals of all nondischarging pacers are then terminated − or, if not immediately terminated, the probability of their termination is at least greatly increased.

The basic idea underlying a model of this general sort is not profound; it is a multi-oscillator system resembling that which is responsible for heartbeat, with the special stipulation that no single element in the system is, by itself, sufficient to trigger (and thereby reset) the entire ensemble. Nevertheless, it would be misleading to suggest that the behavior of the resulting system is transparent or even easy to grasp. Several chapters in this treatise are devoted to exploring the resulting complexity.

The kinds of interaction envisioned here are not necessarily tied to any particular morphological arrangement of components, but the illustration in Figure 4.3 suggests one possible kind of structural arrangement which would be adequate to do the job. The pacers, which are capable of independent, more or less circadian oscillations, provide excitatory input to an integrating element, referred to as a discriminator, during their discharge. As long as the total pacer discharge from the entire ensemble is less than that level specified as threshold, the discriminator is inactive, and the pacers behave in an independent mode. When summed pacer discharge exceeds threshold, the discriminator is activated, and all nondischarging pacers receive a persistent excitatory stimulus which can accelerate their cycles, and lead to premature termination of their recovery intervals.

The manner in which ensemble output, mediated by the discriminator, potentially alters the cycle length of the pacers is a critical aspect of the interaction envisioned (see Fig. 4.6). The basic idea involved follows naturally from an elementary view of the essential dynamics of high-frequency, spontaneously discharging neurons: nerve cells which show tonic responses to steady-state stimuli. The full consequences of this assumption in the present context are far-reaching. Instead of a fully coordinated oscillation like that in the vertebrate heart, a pacemaker of the sort envisioned here develops a very loose-knit kind of mutual entrainment, in which only a fraction of the entire ensemble is truly synchronized. Those pacers which have intrinsic periods shorter than the rhythm developed by the general ensemble will progessively move out of synchrony with the group, their successive cycles scanning through the massed output of the ensemble, until they are eventually engulfed into an earlier cycle than appropriate for synchronization, and this progressive scanning continues indefinitely. Nevertheless, an undamped, self-sustained rhythm develops in the overall behavior of the system, and the cycle-to-cycle precision of the pacemaker output can be orders of magnitude greater than the performance of the single elements of the ensemble. The variance in period of the ensemble has been found to be inversely proportional to the number of pacers participating.

Both the importance of stochastic processes in the system, and the strong nonlinearities which underlie the postulated interactions mean that complete mathematical analysis of this kind of model approaches the intractable. Nevertheless, an adequate appreciation of how a network of this sort functions on a long-term basis can be achieved by computer simulations. In the quantitative formulation necessary for computer simulation of system performance, a set of six parameters have been chosen to describe the ensemble of pacers in their independent, uncoupled state. These parameters serve to quantify: (1) the number of elements or pacers in the ensemble; (2) the overall mean duration of the silent or recovery phase of a "typical" pacer in a "typical" cycle; (3) the cycle-to-cycle stochastic variation in this recovery phase; (4) the variability between pacers of the ensemble in the average duration of the recovery phase; (5) the duration of the active, or discharging portion in the cycle of a typical pacer; and (6) the variability between pacers of the ensemble in the duration of discharge. As dictated by the nature of these parameters, each cycle of a given pacer is that of an unreliable relaxation oscillator, a system of the sort in which something accumulates to a critical stage and then discharges completely. The performance of a pacer varies unpredictably from one cycle to the next, with no serial correlation between cycles. Two additional parameters are required in the models to quantify the coupling process. When a suprathreshold number of pacers in the ensemble are discharging simultaneously, the average recovery phase of each pacer is shortened by a fixed number of hours;

the necessary parameters serve to quantify the threshold, and the amount by which recovery is shortened.

In computer simulations of this sort of system, results have often been obtained in which it was not initially obvious exactly why the model behaves as it does. Because a simulation model can be so easily dissected, however, meaning that the performance of each of its components can be examined in detail, both in isolation and when interacting with the entire network, the internal functioning of such a hypothetical system is open to much more thorough analysis than is at present possible in studies of the real nervous system.

The most interesting parameter of these models is the threshold: that critical value of summed pacer discharge above which interaction among the pacers arises, leading to premature triggering of other, recovering pacers in the ensemble. If one makes the assumption that the value of this threshold depends upon the concurrent light intensity experienced by an animal; and stipulates further that light raises that threshold for nocturnal animals, and lowers threshold for diurnal animals, then a very extensive body of experimental data on the effects of light upon circadian wake-sleep rhythms of birds and mammals can be adequately accounted for in considerable detail. The basic idea expressed by this assumption is that the discriminator, which receives excitatory input from each discharging pacer, is subject to an additional input arising from prevailing light intensity; and that this input constitutes an excitatory stimulus if the animal is diurnally active, and an inhibitory stimulus if the animal is nocturnally active.

The kinds of agreement between behavior of pacemaker models of this sort and the wake-sleep rhythm observed in animal experiments constitute the substance of five chapters in this treatise. One of these chapters deals with responses of the models to different but constant levels of threshold, which are compared with the effects of different intensities of constant light upon the timing of sleep and wakefulness of animals. Correspondences between simulation results and animal experiments include the effects of different intensities of light on free-running period of the rhythm; on duration of wakefulness and of sleep; on cycle-to-cycle variability in times of awakening, times of falling asleep, and the ratio between the two; and on the occurrence of arrhythmicity, without sustained sleep, in birds kept at high levels of constant light.

Another chapter deals with the responses of such pacemaker models to cyclic variations in threshold, and the comparison of these results with the effects of circadian lighting regimes in animal experiments. Correspondences include the timing relationship (i.e., phasing) between an animal's wakefulness and the light cycle during synchronization; the array of differences between the entrainment of diurnal birds and of nocturnal rodents to an equivalent lighting regime; and the striking differences between diurnal and noc-

turnal animals in the manner in which reentrainment to phase-shifted lighting regimes occurs. Of particular interest is the natural way in which a model of this sort, in its most elementary form, can account for synchronization
of the circadian wake-sleep rhythms of rodents, both nocturnal and diurnal,
by lighting regimes with extremely gradual changes in light intensity, including quasi-sinusoidal light cycles of very small amplitude.

Thus the single assumption that light intensity determines the concurrent
value of discriminator threshold in a pacemaker model of this sort provides
a unifying framework for the detailed interpretation of three very broad
classes of experimental results, including, where appropriate, the differences
between the behavior of diurnal and nocturnal animals: observations of
free-running rhythms under constant light intensity; observations of synchronization under square-wave, on-off lighting cycles; and observations of
synchronization under slowly oscillating light regimes.

Two subsequent chapters deal with the responses of such pacemaker models to single brief perturbations of threshold, taken as representative of experiments in which single pulses of light are administered to an animal during an extended free run. In order to reproduce the corresponding experimental results, a critical additional assumption must be incorporated into
the models. In other contexts, the pacemaker was presumed to respond only
to prevailing, concurrent light intensity. The assumption is here required
that a sudden step in light intensity, which sharply increases discriminator
threshold, is accompanied by an overshoot of threshold to a level higher
than at steady state, a transitory supplementary inhibition which persists
for an hour or two. With this additional assumption (which would usually
play a very minor role in synchronization by ordinary light cycles), the
pacemaker models can qualitatively reproduce most of the salient features
of phase shifting induced by single pulses of light in nocturnal rodents
(Chap. 10) and diurnal birds (Chap. 11). These patterns differ markedly in
several ways from each other; corresponding differences in the behavior
of the models arise because of the assumption that light is an inhibitory
stimulus in one case, and an excitatory one in the other.

A broad spectrum of models of the sort considered here can satisfactorily
reproduce the qualitative aspects of experimental data on single-pulse phase
shifting of the circadian wake-sleep rhythm, but proper quantitative agreement between simulation and observation can be attained only by imposing
strong constraints on parameter values, thereby generating a much smaller
subclass of genuinely satisfactory models. For both nocturnal rodents and
diurnal birds, certain aspects of the experimental data seem to demand that
the variability between pacers, in average circadian period when in the uncoupled or independent state, be relatively small. A distribution of pacer
periods with a standard deviation of 1 h gives satisfactory results, but 3 h
is apparently too large; and this is so regardless of how many pacers are in

the ensemble. The extent to which ensemble feedback can accelerate the cycle of a pacer is also constrained by experimental data on phase shifting, with large values (on the order of 12 h) demanded by data from certain nocturnal rodents; and small values (on the order of 4 to 5 h) required for diurnal birds.

When such parameter values are assigned, it is possible for the models to reproduce, in quantitative detail, the phase-shifting patterns observed when single-pulse stimuli are administered to nocturnal rodents and to diurnal birds. A particularly interesting aspect of these simulations is the explanation which they provide for the puzzling experimental observation that phase advance in the rhythm of a nocturnal rodent — but not phase delay — is associated with several so-called transients. After a single phase-advancing stimulus, several circadian cycles are required before steady-state period is reattained in the rhythm, but phase delay is essentially completed within a single cycle. In the models, this turns out to be a natural consequence of the kind of interaction envisioned in the basic formulation. Excitatory coupling, mediated by a threshold, will of necessity leave many of the pacers of the ensemble unentrained; because these unentrained pacers will have periods shorter than that of the ensemble, an asymmetry in the results of phase shifting, like that expressed in transients of the experiments, emerges naturally in the pacemaker models.

Chapter 12 describes what I consider to be the most noteworthy of the correspondences between performance of the models and the behavior observed in animal experiments. Granted an appropriate choice of values for the parameters for the pacemaker, the ensemble can reproduce those changes in free-running period of a circadian rhythm known as after-effects: consistent changes in the period of the wake-sleep rhythm under "standard" conditions, which can be induced by several kinds of pretreatment. Since the average cycle length of an ordinary oscillatory system is one of its most fundamental characteristics, the very existence of such "plasticity" of period in the experimental data could easily be interpreted as evidence for biological processes which go beyond the scope of any elementary attempt to model the system. Several of these kinds of observed after-effects seem superficially to resemble experimental conditioning: the sort of persistent and potentially adaptive response in which the observed rhythm can be "shaped", so that it tends thereafter to show cycles with a period somewhat resembling that which has been externally imposed upon the animal. To find comparable plasticity in a pacemaker model which is as elementary in concept as that proposed here is one of the most interesting phenomena to emerge from the simulations. The essence of the result from computer studies is that a given set of environmental conditions, which affect the threshold of the discriminator, approximately determine the set of average phase relationships among the pacers of the ensemble; and that phase relationships among the pacers can then have long-persistent influences upon the subsequent period of the ensemble, when it is returned to standard conditions.

In most of the experimental data on after-effects, the induced changes of pacemaker period tend gradually to decay, over intervals of weeks to months during constant conditions; the animal seems to "forget". The major component of the after-effects in the models also gradually decays. Certain of the empirical data, however, suggest that after-effects may include a small residual component which represents true change in steady-state period, as though the animal's pacemaker can be brought to alternative states of equilibrium, with different values of period, states which are stable or at least metastable. Some of the results from computer simulations of the model pacemakers similarly suggest that the system can achieve alternative steady states which, in the absence of large perturbation, might persist indefinitely. The evidence on this point from simulation is not compelling — nor are the corresponding experimental data. Hence, a theoretical examination of the model pacemakers assumes particular importance. It can be demonstrated (App. 12.B) that the models *ought,* in principle, to be capable of different states of equilibrium which are stable or metabstable; and that these alternative steady states do not demand unreasonable assumptions about the magnitude of stochastic variation or the properties of the ensemble.

A large portion of this treatise, briefly reviewed in the preceding paragraphs, consists of a demonstration that models of the sort envisioned here are powerful and versatile vehicles for the unified interpretation of a large body of published experimental data on the circadian wake-sleep rhythm. On the basis of these considerations, I would find it surprising indeed if the real pacemakers, which are responsible for the wake-sleep rhythms of the animals, behave as they do as a result of structure and dynamics which are completely different from those envisioned here. Nevertheless, there is not the least doubt in my mind that the models as formulated will prove to be wrong: wrong because they are oversimplified; wrong because they have ignored one critical process or another; and wrong because some elements of the models arise from mistaken points of departure.

The purpose of the predictions described in Chap. 13 is primarily to describe ways in which various possible errors in the models might be exposed. Several predictions are offered which are derived from extremely fundamental aspects of the models, matters which are of central importance to the entire conceptual framework. None of the proposed experiments is technically impossible, and most of them would, I think, be relatively easy to perform. Should the predictions for these experiments prove to be mistaken, an attempt to rescue the models from complete rejection would be difficult to justify, since there would probably be so little left of their original substance.

Other predictions are described involving features of the models which are not absolutely essential to the basic concepts, but only appealingly simple ways in which such a system conceivably might function. Should these other predictions prove to be mistaken, some sort of rescue operation could be undertaken, leading to a restructured and very probably less simple kind of model. While one cannot fully anticipate the direction these modifications might taken, it is my guess that such revisions would lead to alternative models which are less attractive than that initially proposed.

The third kind of prediction included in Chap. 13 deals with detailed quantitative use of the models. The preliminary requirement for such prediction is that parameters of the appropriate specific model have been adequately and completely evaluated. For illustrative purposes, rough assumptions about these values are made; the resulting quantitative predictions are therefore not intended to define expectations for any given species of animal, but instead only to indicate the kinds of diverse use to which a completely evaluated model can be put.

Further predictions are described in Chap. 14, the first of which involve a novel experimental protocol, in which timing of a repetitive light stimulus is linked to the animal's rhythmic behavior (clamped free run). That technique can be used effectively to evaluate the most characteristic feature of the way this class of models responds to lighting: the influence upon discriminator threshold (and thereby upon period of the rhythm) of that light intensity which prevails at the time of awakening. Results from a preliminary application of this experimental design are offered which demonstrate its feasibility.

Two problem cases are then described: sets of existing experimental data which do not conform with predictions which follow naturally from the elementary form of the models. To reconcile the first of these instances with the models demands an additional parameter which specifies interpacer variability in the effects of discriminator feedback on the pacers; and requires, in addition, the elimination of that correlation, between period of a pacer and duration of its discharge, which resulted from the equations initially chosen for the models. The second kind of problematic data involves the occasional cases in which an animal's activity pattern splits into two "components", which can then free-run with different periods. The pacemaker models here, in their elementary form, do not have that capacity. To account for some of the details in such data in the present context would require a pacemaker system with two semiautonomous discriminators, as well as pacers which receive input from one discriminator and which have output to the other. The kind of additional assumptions required by these data is a clear demonstration that the circadian pacemakers of some species are appreciably more complex than I have assumed.

Predictions of a qualitatively different sort are described in Chap. 15. Interest there is focussed upon how the models might map upon physiological substrate, and what clues are available to the experimentalist, by which he might recognize morphological equivalents of the model pacemakers and their components. In order to distinguish between the circadian pacemaker and some secondary physiological process driven by the pacemaker, advantage might be taken of certain unusual experimental situations in which the pacemaker can be presumed — retrospectively — to have been functioning normally, while its ordinary output (the wake-sleep, activity-rest cycle) was suppressed. Two opportunities for this sort of investigation are described; the appropriate experimental investigations in these cases seem entirely feasible. More difficult are the experimental techniques required to distinguish between the kind of pacemaker envisioned here, and one which operates by entirely different principles. For this, it would be essential to monitor in the intact animal the behavior of single cells of the presumptive pacemaker over several days. One of the critical predictions from the models is that many such cells should have an average period shorter than that of the pacemaker ensemble. These would represent the unentrained, short-period pacers, which were essential to the manner in which the models accounted for several important features of the rhythm data.

The bulk of Chap. 15 is devoted to an interpretation of an extensive series of experiments on the rhythms of birds in terms of the pacemaker models envisioned here. Those experiments demonstrate that the pineal organ of some species of birds plays an important role in their circadian rhythmicity. The results can be interpreted by the hypothesis that the pineal organ participates in the coupling among oscillatory neurons located elsewhere; as such, it may correspond to the discriminator — or at least to a portion of the discriminator — in the pacemakers of this treatise. A series of predictions is therefore based on that interpretation; and these predictions provide a means by which the postulated correspondence can be further tested.

Perhaps the most appropriate and useful way to close this summary chapter is with a restatement of the two interpretations which I would select as the most noteworthy:

1. A biological pacemaker which is as extremely precise as the best of circadian rhythms could readily be assembled out of a relatively modest number of sloppy components: elements which are only more or less circadian in period, which are diverse in their individual properties and which are unreliable in their cycle-to-cycle performance. An adequate form of coupling to accomplish this can be derived from simple excitatory interactions, using properties and behavior similar to those of high-frequency neurons. No other mechanistic interpretation for the precision of circadian rhythms is currently available.

2. A pacemaker consisting of an ensemble of relaxation oscillators, which are coupled in the manner envisioned here, can account naturally for experimental data on after-effects (small, inducible, and long-persistent changes in period of a circadian rhythm). The storage of information implicit in the after-effects could reside in a dynamic memory, inherent only in the phase relationships among oscillators. The most conspicuous of after-effects, both in the experimental data and in the models, are transitory; they eventually decay. Under certain moderately restrictive conditions, however, induced changes in the phasing of pacers in the models should allow the pacemaker to enter alternative steady states, with slightly different values of average period. These states could each be stable, i.e., self-perpetuating, in spite of appreciable stochastic noise within the system. Transitory after-effects could also be accounted for by weak mutual coupling between only two oscillators (cf. Daan and Berde 1978), but no other mechanistic interpretation is currently available for an array of alternative *steady-state* values of period in a circadian pacemaker.

References

Aschoff J (1958) Tierische Periodik unter dem Einfluß von Zeitgebern. Z Tierpsychol 15: 1–30

Aschoff J (1961) Exogenous and endogenous components in circadian rhythms. Cold Spring Harbor Symp Quant Biol 25: 11–28

Aschoff J (1965a) Circadian rhythms in man. Science 148: 1427–1432

Aschoff J (1965b) Response curves in circadian periodicity. In: Aschoff J (ed) Circadian clocks. North-Holland Publishing Company, Amsterdam, pp 95–111

Aschoff J (1979) Circadian rhythms: influences of internal and external factors on the period measured in constant conditions. Z Tierpsychol in press

Aschoff J, Diehl I, Gerecke U, Wever R (1962) Aktivitätsperiodik von Buchfinken (*Fringilla coelebs* L.) unter konstanten Bedingungen. Z Vergl Physiol 45: 605–617

Aschoff J, Gerecke U, Kureck A, Pohl H, Rieger P, Saint Paul U, Wever R (1971) Interdependent parameters of circadian activity rhythms in birds and man. In: Menaker M (ed) Biochronometry. Nat Acad Sci, Washington DC, pp 3–27

Audesirk G, Strumwasser F (1975) Circadian rhythm of neuron R-15 of *Aplysia californica:* in vivo photoentrainment. Proc Natl Acad Sci USA 72: 2408–2412

Ayers JL, Selverston AI (1979) Monosynaptic entrainment of an endogenous pacemaker network: a cellular mechanism for von Holst's magnet effect. J Comp Physiol 129: 5–18

Barlow JS (1961) Discussion. Cold Spring Harbor Symp Quant Biol 25: 54–55

Beiswanger CM, Jacklet JW (1975) In vitro tests for a circadian rhythm in the electrical activity of a single neuron in *Aplysia californica*. J Comp Physiol 103: 19–37

Binkley S, Kluth E, Menaker M (1972) Pineal and locomotor activity: levels and arrhythmia in sparrows. J Comp Physiol 77: 163–169

Binkley SA, Riebman JB, Reilly KB (1978) The pineal gland: a biological clock in vitro. Science 202: 1198–1200

Borbély AA, Neuhaus HU (1978a) Daily pattern of sleep, motor activity and feeding in the rat: effects of regular and gradually extended photoperiods. J Comp Physiol 124: 1–14

Borbély AA, Neuhaus HU (1978b) Circadian rhythm of sleep and motor activity in the rat during skeleton photoperiod, continuous darkness and continuous light. J Comp Physiol 128: 37–46

Bruce VG (1961) Environmental entrainment of circadian rhythms. Cold Spring Harbor Symp Quant Biol 25: 29–47

Buck J, Buck E (1968) Mechanism of rhythmic synchronous flashing of fireflies. Science 159: 1319–1327

Buller AJ (1965) A model illustrating some aspects of muscle spindle physiology. J Physiol 179: 402–414

Bullock TH, Hamstra RH, Scheich H (1972) The jamming avoidance response of high frequency electric fish. J Comp Physiol 77: 1–48

Calvin WH, Stevens CF (1965) A Markov process model for neuron behavior in the interspike interval. Proc 18th Ann Conf Engineering in Med Biol, p 118

Coindet J, Chouvet G, Mouret J (1975) Effects of lesions of the suprachiasmatic nuclei on paradoxical sleep and slow wave sleep circadian rhythms in the rat. Neurosci Lett 1: 243–247

Crowley T, Kripke D, Halberg F, Pegram G, Schildkraut J (1972) Circadian rhythms of
  Macaca mulatta: sleep, EEG, body and eye movement and temperature. Primates 13:
  149–168
Daan S, Berde C (1978) Two coupled oscillators: simulations of the circadian pacemaker
  in mammalian activity rhythms. J Theor Biol 70: 297–313
Daan S, Pittendrigh CS (1976a) A functional analysis of circadian pacemakers in noc-
  turnal rodents. II. The variability of phase response curves. J Comp Physiol 106:
  253–266
Daan S, Pittendrigh CS (1976b) A functional analysis of circadian pacemakers in noc-
  turnal rodents. III. Heavy water and constant light: homeostasis of frequency? J
  Comp Physiol 106: 367–290
DeCoursey PJ (1960) Daily rhythm of light sensitivity in a rodent. Science 131: 33–35
DeCoursey PJ (1961a) Effect of light on the circadian rhythm of the flying squirrel,
  Glaucomys volans. Z Vergl Physiol 44: 331–354
DeCoursey PJ (1961b) Phase control of activity in a rodent. Cold Spring Harbor Symp
  Quant Biol 25: 49–54
DeCoursey PJ (1964) Function of a light response rhythm in hamsters. J Cell Comp
  Physiol 63: 189–196
Deguchi T (1979) Circadian rhythm of serotonin N-Acetyltransferase activity in organ
  culture of chicken pineal gland. Science 203: 1245–1247
Enright JT (1965) Synchronization and ranges of entrainment. In: Aschoff J (ed) Cir-
  cadian clocks. North-Holland Publishing Company, Amsterdam, pp 112–124
Enright JT (1966a) Influence of seasonal factors on the activity onset of the house
  finch. Ecology 47: 662–666
Enright JT (1966b) Temperature and the free-running circadian rhythm of the house
  finch. Comp Biochem Physiol 18: 463–475
Enright JT (1967) The spontaneous neuron subject to tonic stimulation. J Theor Biol
  16: 54–77
Enright JT (1972) A virtuoso isopod. Circa-lunar rhythms and their tidal fine structure.
  J Comp Physiol 77: 141–162
Enright JT (1975) The circadian tape recorder and its entrainment. In: Vernberg J (ed)
  Physiological adaptation to the environment. Intext Educ Publ, New York, pp 465–
  476
Erkert HG (1969) Die Bedeutung des Lichtsinnes für Aktivität und Raumorientierung
  der Schleiereule (*Tyto alba guttata* Brehm). Z Vergl Physiol 64: 37–70
Eskin A (1971) Some properties of the system controlling the circadian activity rhythm
  of sparrows. In: Menaker M (ed) Biochronometry. Nat Acad Sci, Washington DC,
  pp 55–78
Fohlmeister J, Peppele RE, Purple RL (1974) Repetitive firing: dynamic behavior of
  sensory neurons reconciled with a quantitative model. J Neurophysiol 37: 1213–1227
French AR (1977) Periodicity of recurrent hypothermia during hibernation in the pocket
  mouse, *Perognathus longimembris.* J Comp Physiol 115: 87–100
Gaston S (1971) The influence of the pineal organ on the circadian activity rhythms
  in birds. In: Menaker J (ed) Biochronometry. Nat Acad Sci, Washington DC, pp
  541–549
Gaston S, Menaker M (1968) Pineal function: the biological clock in the sparrow? Science
  160: 1125–1127
Geisler CD, Goldberg JM (1966) A stochastic model of the repetitive activity of neurons.
  Biophys J 6: 53–69
Gwinner E (1974) Testosterone induces "splitting" of circadian locomotor activity
  rhythms in birds. Science 185: 72–74
Gwinner E (1978) Effects of pinealectomy on circadian locomotor activity rhythms in
  European starlings, *Sturnus vulgaris.* J Comp Physiol 126: 123–129
Gwinner E, Benzinger I (1978) Synchronization of a circadian rhythm in pinealectomized
  European starlings by daily injection of melatonin. J Comp Physiol 127: 209–214

Hagiwara S (1954) Analysis of interval fluctuation of the sensory nerve impulse. Jpn
    J Physiol 4: 234–240
Hamner WM, Enright JT (1967) Relationships between photoperiodism and circadian
    rhythms of activity in the house finch. J Exp Biol 46: 43–61
Hartline DK (1976) Simulation of phase-dependent pattern changes to perturbations
    of regular firing in crayfish stretch receptors. Brain Res 110: 245–257
Hoffmann K (1965) Overt circadian frequencies and circadian rule. In: Aschoff J (ed)
    Circadian clocks. North-Holland Publishing Company, Amsterdam, pp 87–94
Hoffmann K (1971) Splitting of the circadian rhythm as a function of light intensity.
    In: Menaker M (ed) Biochronometry. Nat Acad Sci, Washington DC, pp 134–146
Holst von E (1939) Die relative Koordination als Phänomen und als Methode zentralner-
    vöser Funktionsanalyse. Ergeb Physiol 42: 228–306
Karakashian MW, Schweiger HG (1976) Circadian properties of the rhythmic system in
    individual nucleated and enucleated cells of *Acetabularia mediterranea*. Exp Cell Res
    97: 366–377
Kasal CA, Menaker M, Perez-Polo JR (1979) Circadian clock in culture: N-acetyltrans-
    ferase activity of chick pineal glands oscillates in vitro. Science 203: 656–658
Kleitman N (1923) The effects of prolonged sleeplessness on man. Am J Physiol 66:
    67–92
Kleitman N (1963) Sleep and wakefulness. Univ Chicago Press, Chicago
Kramm KR (1971) Circadian activity in the antelope ground squirrel, *Ammospermo-
    philus leucurus*. Ph D Diss Univ Calif Irvine
Lickey ME (1966) Further studies of a circadian rhythm in a single neuron: seasonal
    modulation and adiurnal entrainment. Physiologist 9: 230
Lickey ME (1969) Seasonal modulation and non-24-hour entrainment of a circadian
    rhythm in a single neuron. J Comp Physiol Psychol 68: 9–17
Lickey ME, Zack S, Birrell P (1971) Some factors governing entrainment of a circadian
    rhythm in a single neuron. In: Menaker M (ed) Biochronometry. Nat Acad Sci,
    Washington DC, pp 549–560
Lindberg RG, Hayden P (1974) Research on the properties of circadian systems amenable
    to study in space. Final report. Northrop Research and Technology Center, Haw-
    thorne Calif
Lindberg RG, Gambino JJ, Hayden P (1971) Circadian periodicity of resistance to ioniz-
    ing radiation in the pocket mouse. In: Menaker M (ed) Biochronometry. Nat Acad
    Sci, Washington DC, pp 169–184
Martinez JL (1972) Effects of selected illumination levels on circadian periodicity in
    the rhesus monkey (*Macaca mulatta*). J Interdiscip Cycle Res 1: 47–59
McMillan JP (1972) Pinealectomy abolishes the circadian rhythm of migratory restless-
    ness. J Comp Physiol 79: 105–112
McNew J, Burson R, Hoshizaki T, Adey R (1972) Sleep-wake cycle of an unrestrained
    chimpanzee under entrained and free-running conditions. Aerosp Med 43: 155–161
Menaker M (1968) Extraretinal light perception in the sparrow. I. Entrainment of the
    biological clock. Proc Natl Acad Sci USA 59: 414–421
Menaker M (1971) Synchronization with the photic environment via extraretinal re-
    ceptors in the avian brain. In: Menaker M (ed) Biochronometry. Nat Acad Sci, Wash-
    ington DC, pp 315–332
Menaker M, Zimmerman N (1976) Role of the pineal in the circadian system of birds.
    Am Zool 16: 45–55
Mittler MM, Lund R, Sokolove PG, Pittendrigh CS, Dement WC (1977) Sleep and activ-
    ity rhythms in mice: a description of circadian patterns and unexpected disruptions
    in sleep. Brain Res 131: 129–145
Parzen E (1960) Modern probability theory and its applications. John Wiley and Sons,
    New York
Pavlidis T (1969) Populations of interacting oscillators and circadian rhythms. J Theor
    Biol 22: 418–436

Pavlidis T (1971) Populations of biochemical oscillators as circadian clocks. J Theor
    Biol 33: 319–338
Pavlidis T (1973) Biological oscillators: their mathematical analysis. Academic Press,
    London New York, 207 pp
Pelham RW (1975) A serum melatonin rhythm in chickens and its abolition by pineal-
    ectomy. Endocrinology 96: 543–546
Piéron H (1913) Le probleme physiologique du sommeil. Masson et Cie, Paris, 520 pp
Pittendrigh CS (1961) Circadian rhythms and the circadian organization of living sys-
    tems. Cold Spring Harbor Symp Quant Biol 25: 159–181
Pittendrigh CS (1965) On the mechanism of the entrainment of a circadian rhythm by
    light cycles. In: Aschoff J (ed) Circadian clocks. North-Holland Publishing Company,
    Amsterdam, pp 277–297
Pittendrigh CS (1967) Circadian rhythms, space research and manned space flight. In:
    Life sciences and space research V. North-Holland Publishing Company, Amsterdam
Pittendrigh CS, Daan S (1976a) A functional analysis of circadian pacemakers in noc-
    turnal rodents. I. Stability and lability of spontaneous frequency. J Comp Physiol
    106: 223–252
Pittendrigh CS, Daan S (1976b) A functional analysis of circadian pacemakers in noc-
    turnal rodents. IV. Entrainment: Pacemaker as clock. J Comp Physiol 106: 291–333
Platt JR (1964) Strong inference. Science 146: 347–353
Pohl H (1972) Die Aktivitätsperiodik von zwei tagaktiven Nagern, *Funambulus palmarum*
    and *Eutamius sibericus,* unter Dauerlichtbedingungen. J Comp Physiol 78: 60–74
Rutledge JT, Angle MJ (1977) Persistence of circadian activity rhythms in pinealecto-
    mized European starlings (*Sturnus vulgaris*). J Exp Biol 202: 333–338
Simpson SM, Follett BK (1979) The role of the pineal and the anterior hypothalamus
    in regulating the circadian rhythms of locomotor activity in the Japanese quail. Proc.
    XVIIth Int Ornithol Congr, Berlin, in press
Strumwasser F (1965) The demonstration and manipulation of a circadian rhythm in
    a single neuron. In: Aschoff J (ed) Circadian clocks. North-Holland Publishing Com-
    pany, Amsterdam, pp 442–462
Swade RH (1969) Circadian rhythms in fluctuating light cycles: toward a new model
    of entrainment. J Theor Biol 24: 227–239
Swade RH, Pittendrigh CS (1967) Circadian locomotor rhythms of rodents in the arctic.
    Am Nat 101: 431–466
Turek FW, McMillan JP, Menaker M (1976) Melatonin: effects on the circadian loco-
    motor rhythm of sparrows. Science 194: 1441–1443
Underwood H (1977) Circadian organization in lizards: the role of the pineal organ.
    Science 195: 587–589
Verveen AA, Derksen HE (1965) Fluctuations in membrane potential of axons and the
    problem of coding. Kybernetik 2: 152–160
Viernstein LJ, Grossman RG (1961) Neural discharge patterns in the transmission of
    sensory information. In: Cherry C (ed) Information theory, 4th London Symp,
    Butterworths, London
Wahlström G (1964) The circadian rhythm in the canary studied by self-selection of
    light and darkness. Acta Soc Med Upsala 69: 241–271
Wahlström G (1965) The circadian rhythm of self-selected rest and activity in the canary
    and the effects of barbiturates, reserpine, monoamine oxidase inhibitors and enforced
    darkness periods. Acta Physiol Scand 65: Suppl 250: 1–67
Weiss TF (1964) A model for firing patterns of auditory nerve fibers. Tech Rept Mass
    Inst Technol Res Lab Electron No 418, 95 pp
Wever R (1962) Zum Mechanismus der biologischen 24-Stunden-Periodik. Kybernetic
    1: 139–154
Wever R (1963) Zum Mechanismus der biologischen 24-Stunden-Periodik. Kybernetic
    1: 213–228
Wever R (1964) Zum Mechanismus der biologischen 24-Stunden-Periodik. Kybernetik
    2: 127–144

Wiener N (1958) Nonlinear problems in random theory. John Wiley and Sons, Cambridge Mass

Winfree AT (1967) Biological rhythms and the behavior of populations of coupled oscillators. J Theor Biol 16: 15–42

Zimmerman N (1976) Organization within the circadian system of the house sparrow: hormonal coupling and the location of a circadian oscillator. Ph D Diss Univ Texas Austin

Zimmerman N, Menaker M (1975) Neural connections of sparrow pineal: role in circadian control of activity. Science 190: 477–479

Author Index

Subject Index

Studies of Brain Function

Coordinating Editor: V. Braitenberg
Editors: H. B. Barlow, E. Florey, O.-J. Grüsser,
H. van der Loos

"With this small volume a new series begins, devoted to a fascinating theme, *i. e.* the brain function. As stated in the preface attention is centered on the functional aspects of brains, and this will remain a constant feature of the whole series. In this first booklet the relation between neural activity and behavior is particulary emphasized. Moreover, the author points out how, among vertebrates, many species of fish may be considered of particular interest from the standpoint of neuroethological research, since behavioral responses are very often clearly related to neuronal events. ...The concise and clear style of the book together with the high (58) number of figures, which allows for a better understanding of the matter, will be greatly appreciated by the interested reader to whom the reviewer recommends it as a precious introduction in the field."

Electroanalytical Abstracts

"...the author of this monograph has provided sufficient background material to enable most serious neurophysiologists to benefit from a reading of the book. Most of the information is concerned with experiments on the vestibular apparatus of two species, cat and frog. ... A great deal of attention is paid the physiology of the vestibular nuclei, to vestibulo-cerebellar relationships and to vestibulo-ocular relationships. This is a field to which the author has been careful and tireless contributor.
The book is written in a compact style, and an impressive amount of material has been incorporated into a rather small book. The book encompasses a surprising breadth of literature. It is far more than the simple account of the author's work than it might appear to be from the introduction. Much of the current literature and much of the older work in this field is considered and discussed."

Trends in Neuroscience

Springer-Verlag
Berlin
Heidelberg
New York

Behavioral Ecology and Sociobiology

Managing Editor: H. Markl
in cooperation with a distinguished advisory board

Behavioral Ecology and Sociobiology was founded by Springer-Verlag in 1976 as an international journal. Drawing on a philosophy developed and nurtured for more than half a century in the original *Zeitschrift für vergleichende Physiologie* (now *Journal of Comparative Physiology*), it presents original articles and short communications dealing with the experimental analysis of animal behavior on an individual level and in population. Special emphasis is given to the functions, mechanisms, and evolution of ecological adaptations of behavior. Specific areas covered include:

- orientation in space and time
- communication and all other forms of social and interspecific behavior
- origins and mechanisms of behavior preferences and aversions, e.g., with respect to food, locality, and social partners
- behavioral mechanisms of competition and resource partitioning
- population physiology
- evolutionary theory of social behavior.

Behavioral Ecology and Sociobiology is designed to serve as a link between researchers and students in a variety of disciplines.

Subscription Information upon request.

Journal of Comparative Physiology · A+B

Founded in 1924 as
Zeitschrift für vergleichende Physiologie
by K. von Frisch and A. Kühn

A. Sensory, Neural, and Behavioral Physiology

Editorial Board: H. Autrum, R. R. Capranica, C. Delcomyn, K. von Frisch, G. A. Horridge, M. Lindauer, C. L. Prosser

Advisory Board: H. Atwood, S. Daan, W. H. Fahrenbach, B. Hölldobler, Y. Katsuki, M. Konishi, M. F. Land, M. S. Laverack, H. C. Lüttgau, H. Markl, A. Michelsen, D. Ottoson, F. Papi, C. S. Pittendrigh, W. Precht, J. D. Pye, A. Roth, H. F. Rowell, D. G. Stavenga, R. Wehner, J. J. Wine

B. Biochemical, Systemic, and Environmental Physiology

Editorial Board: K. Johansen, B. Linzen, W. T. W. Potts, C. L. Prosser

Advisory Board: G. A. Bartholomew, H. Bern, P. J. Butler, Th. Eisner, D. H. Evans, S. Nilsson, O. Randall, R. B. Reeves, G. H. Satchell, T. J. Shuttleworth, G. Somero, K. Urich, S. Utida, G. R. Wyatt, E. Zebe

The *Journal of Comparative Physiology* publishes original articles in the field of animal physiology. In view of the increasing number of papers and the high degree of scientific specialization the journal is published in two sections.

A. Sensory, Neural, and Behavioral Physiology
Physiological Basis of Behavior; Sensory Physiology; Neural Physiology; Orientation, Communication; Locomotion; Hormonal Control of Behavior

B. Biochemical, Systemic, and Environmental Physiology
Comparative Aspects of Metabolism and Enzymology; Metabolic Regulation; Respiration and Gas Transport; Physiology of Body Fluids; Circulation; Temperature Relations; Muscular Physiology

Subscription Information upon request.

Springer-Verlag
Berlin
Heidelberg
New York